Reviews of Environmental Contamination and Toxicology

VOLUME 148

Springer
New York
Berlin
Heidelberg
Barcelona
Budapest
Hong Kong
London
Milan
Paris
Santa Clara
Singapore
Tokyo

Reviews of Environmental Contamination and Toxicology

Continuation of Residue Reviews

Editor
George W. Ware

VOLUME 148

Springer

Springer-Verlag
New York: 175 Fifth Avenue, New York, NY 10010, USA
Heidelberg: 69042 Heidelberg, Postfach 10 52 80, Germany

Library of Congress Catalog Card Number 62-18595.
Printed in the United States of America.

ISSN 0179-5953

Printed on acid-free paper.

ISBN 0-387-94842-2 Springer-Verlag New York Berlin Heidelberg   SPIN 10523393

Foreword

International concern in scientific, industrial, and governmental communities over traces of xenobiotics in foods and in both abiotic and biotic environments has justified the present triumvirate of specialized publications in this field: comprehensive reviews, rapidly published research papers and progress reports, and archival documentations. These three international publications are integrated and scheduled to provide the coherency essential for nonduplicative and current progress in a field as dynamic and complex as environmental contamination and toxicology. This series is reserved exclusively for the diversified literature on "toxic" chemicals in our food, our feeds, our homes, recreational and working surroundings, our domestic animals, our wildlife and ourselves. Tremendous efforts worldwide have been mobilized to evaluate the nature, presence, magnitude, fate, and toxicology of the chemicals loosed upon the earth. Among the sequelae of this broad new emphasis is an undeniable need for an articulated set of authoritative publications, where one can find the latest important world literature produced by these emerging areas of science together with documentation of pertinent ancillary legislation.

Research directors and legislative or administrative advisers do not have the time to scan the escalating number of technical publications that may contain articles important to current responsibility. Rather, these individuals need the background provided by detailed reviews and the assurance that the latest information is made available to them, all with minimal literature searching. Similarly, the scientist assigned or attracted to a new problem is required to glean all literature pertinent to the task, to publish new developments or important new experimental details quickly, to inform others of findings that might alter their own efforts, and eventually to publish all his/her supporting data and conclusions for archival purposes.

In the fields of environmental contamination and toxicology, the sum of these concerns and responsibilities is decisively addressed by the uniform, encompassing, and timely publication format of the Springer-Verlag (Heidelberg and New York) triumvirate:

Reviews of Environmental Contamination and Toxicology [Vol. 1 through 97 (1962–1986) as Residue Reviews] for detailed review articles concerned with any aspects of chemical contaminants, including pesticides, in the total environment with toxicological considerations and consequences.

Bulletin of Environmental Contamination and Toxicology (Vol. 1 in 1966)
for rapid publication of short reports of significant advances and discoveries in the fields of air, soil, water, and food contamination and pollution as well as methodology and other disciplines concerned with the introduction, presence, and effects of toxicants in the total environment.

Archives of Environmental Contamination and Toxicology (Vol. 1 in 1973)
for important complete articles emphasizing and describing original experimental or theoretical research work pertaining to the scientific aspects of chemical contaminants in the environment.

Manuscripts for *Reviews* and the *Archives* are in identical formats and are peer reviewed by scientists in the field for adequacy and value; manuscripts for the *Bulletin* are also reviewed, but are published by photo-offset from camera-ready copy to provide the latest results with minimum delay. The individual editors of these three publications comprise the joint Coordinating Board of Editors with referral within the Board of manuscripts submitted to one publication but deemed by major emphasis or length more suitable for one of the others.

Coordinating Board of Editors

Preface

Thanks to our news media, today's lay person may be familiar with such environmental topics as ozone depletion, global warming, greenhouse effect, nuclear and toxic waste disposal, massive marine oil spills, acid rain resulting from atmospheric SO_2 and NO_x, contamination of the marine commons, deforestation, radioactive leaks from nuclear power generators, free chlorine and CFC (chlorofluorocarbon) effects on the ozone layer, mad cow disease, pesticide residues in foods, green chemistry or green technology, volatile organic compounds (VOCs), hormone- or endocrine-disrupting chemicals, declining sperm counts, and immune system suppression by pesticides, just to cite a few. Some of the more current, and perhaps less familiar, additions include *xenobiotic transport, solute transport, Tiers 1 and 2, USEPA to cabinet status, and zero-discharge*. These are only the most prevalent topics of national interest. In more localized settings, residents are faced with leaking underground fuel tanks, movement of nitrates and industrial solvents into groundwater, air pollution and "stay-indoors" alerts in our major cities, radon seepage into homes, poor indoor air quality, chemical spills from overturned railroad tank cars, suspected health effects from living near high-voltage transmission lines, and food contamination by "flesh-eating" bacteria and other fungal or bacterial toxins.

It should then come as no surprise that the '90s generation is the first of mankind to have become afflicted with *chemophobia*, the pervasive and acute fear of chemicals.

There is abundant evidence, however, that virtually all organic chemicals are degraded or dissipated in our not-so-fragile environment, despite efforts by environmental ethicists and the media to persuade us otherwise. However, for most scientists involved in environmental contaminant reduction, there is indeed room for improvement in all spheres.

Environmentalism is the newest global political force, resulting in the emergence of multi-national consortia to control pollution and the evolution of the environmental ethic. Will the new politics of the 21st century be a consortium of technologists and environmentalists or a progressive confrontation? These matters are of genuine concern to governmental agencies and legislative bodies around the world, for many serious chemical incidents have resulted from accidents and improper use.

For those who make the decisions about how our planet is managed, there is an ongoing need for continual surveillance and intelligent controls to avoid endangering the environment, the public health, and wildlife. Ensuring safety-in-use of the many chemicals involved in our highly industrialized culture is a dynamic challenge, for the old, established materials are continually being displaced by newly developed molecules more acceptable to federal and state regulatory agencies, public health officials, and environmentalists.

Adequate safety-in-use evaluations of all chemicals persistent in our air, foodstuffs, and drinking water are not simple matters, and they incorporate the judgments of many individuals highly trained in a variety of complex biological, chemical, food technological, medical, pharmacological, and toxicological disciplines.

Reviews of Environmental Contamination and Toxicology continues to serve as an integrating factor both in focusing attention on those matters requiring further study and in collating for variously trained readers current knowledge in specific important areas involved with chemical contaminants in the total environment. Previous volumes of *Reviews* illustrate these objectives.

Because manuscripts are published in the order in which they are received in final form, it may seem that some important aspects of analytical chemistry, bioaccumulation, biochemistry, human and animal medicine, legislation, pharmacology, physiology, regulation, and toxicology have been neglected at times. However, these apparent omissions are recognized, and pertinent manuscripts are in preparation. The field is so very large and the interests in it are so varied that the Editor and the Editorial Board earnestly solicit authors and suggestions of underrepresented topics to make this international book series yet more useful and worthwhile.

Reviews of Environmental Contamination and Toxicology attempts to provide concise, critical reviews of timely advances, philosophy, and significant areas of accomplished or needed endeavor in the total field of xenobiotics in any segment of the environment, as well as toxicological implications. These reviews can be either general or specific, but properly they may lie in the domains of analytical chemistry and its methodology, biochemistry, human and animal medicine, legislation, pharmacology, physiology, regulation, and toxicology. Certain affairs in food technology concerned specifically with pesticide and other food-additive problems are also appropriate subjects.

Justification for the preparation of any review for this book series is that it deals with some aspect of the many real problems arising from the presence of any foreign chemical in our surroundings. Thus, manuscripts may encompass case studies from any country. Added plant or animal pest-control chemicals or their metabolites that may persist into food and animal feeds are within this scope. Food additives (substances deliberately added to foods for flavor, odor, appearance, and preservation, as well as those inadvertently added during manufacture, packing, distribution, and storage) are also considered suitable review material. Additionally, chemical contamination in any manner of air, water, soil, or plant or animal life is within these objectives and their purview.

Normally, manuscripts are contributed by invitation, but suggested topics are welcome. Preliminary communication with the Editor is recommended before volunteered review manuscripts are submitted.

Department of Entomology   G.W.W.
University of Arizona
Tucson, Arizona

Table of Contents

Toxicity of Nickel to Soil Organisms in Denmark

Janeck J. Scott-Fordsmand*

Contents

I. Introduction

Nickel (Ni) is a naturally occurring element that is present in soil, water, air, and biological material. Although Ni occurs naturally, concentrations found in the environment may also be caused by anthropogenic input such as deposition from the burning of fossil fuels. Merian (1984) estimated the global annual anthropogenic release to the environment to be approximately 180,000 tonnes, in addition to the 150,000 tonnes that is redistributed by natural processes.

Although Ni is omnipresent and is vital for the function of many organisms, concentrations in some areas from both anthropogenic release and naturally varying levels may be toxic to living organisms. In humans, Ni is known to cause liver, kidney, spleen, and brain damage on acute exposure and vesicular eczema, lung and nasal cancer, and tissue damage upon

---

* National Environmental Research Institute, Department of Terrestrial Ecology, P.O. Box 314, Vejlsøvej 25, DK-8600 Silkeborg, Denmark.

© 1997 by Springer-Verlag New York, Inc.
Reviews of Environmental Contamination and Toxicology, Vol. 148.

chronic exposure (IPCS 1992). For other organisms, a variety of effects have been observed at different environmental levels. For microorganisms, Ni is known to inhibit growth and some enzymatic functions. In plants, effects such as patchy discoloration, premature senescence, yellowing of old leaves, and growth reduction may be observed; in animals, tissue damage, reduced reproduction or growth, and mortality have been observed.

This review is concerned with the ecotoxicological effects of Ni in the terrestrial environment, especially effects upon soil-dwelling organisms. Only data from observations in the soil media have been considered; plant studies that were performed in solution cultures or with microorganisms grown on plates are not included. After a short summary of the sources, fate, and bioavailability of Ni in soil, with emphasis on the Danish situation, the major part of this paper presents ecotoxicological data of Ni concerned with soil-dwelling microorganisms, plants, and invertebrates. The data were collected from laboratory and field experiments published in international papers and reports. Finally, a short evaluation of the ecological risks of Ni to the terrestrial environment, with Denmark as an example, and risk assessments from selected countries are presented.

II. Chemical/Physical Properties

Nickel is a metallic element belonging to group VIIIb of the periodic table. It is silvery white, hard, malleable, ductile, and a good conductor of heat and electricity. It is found in several oxidation states and is insoluble in water, soluble in dilute nitric acid (HNO_3), slightly soluble in hydrochloric acid (HCl) and sulfuric acid (H_2SO_4), and insoluble in ammonium hydroxide (NH_4OH) (Table 1). In biological systems, nickel in its ionic form may form complexes and bind to organic material, and in some organisms it is an essential component of certain molecules. The prevalent ionic form is Ni^{2+} (Table 1).

Table 1. Chemical and physical properties of nickel.

Nickel	Physical/chemical data
Atomic number	28
Atomic weight	58.71 g mol^{-1}
Melting point	1453 °C
Boiling point	2732 °C
Density	8.90 g cm^3
Solubility (water)	Insoluble
Oxidation state	$0, +1, +2, +3, +4, +5$
Isotopes	^{58}Ni (68%), ^{60}Ni (26%), ^{61}Ni (1%), ^{62}Ni (4%), ^{64}Ni (4%)
CAS No.	7440-02-0

III. Sources and Usage

Nickel is a natural component of Earth's crust, is present in igneous rocks, and is ubiquitous in the environment. There are almost 100 minerals of which Ni is an essential constituent. The primary Ni ore is laterite, a nickel oxide ore mined mostly by open-pit techniques in Australia, Cuba, Indonesia, New Caledonia, the United States, and the former Soviet Union (Adriano 1986). Underground, the primary ore mined is pentlandite. In Denmark, the major uses of nickel are in alloy production (69%), in the electroplating industry (17%), and in the production of Ni–Cd batteries (Miljøstyrelsen 1985). Worldwide, the production of stainless steel accounts for a large proportion of the Ni used. Nickel is also used for the production of other electronic components and in the preparation of catalysts for hydrogenation of fats and methanation.

IV. Background Concentrations in Soil

Adriano (1986) estimated the worldwide average soil concentration to be about 40 mg Ni kg^{-1}, but the concentration may vary from less than 1 mg Ni kg^{-1} to several thousand milligrams per kilogram. A low content is normally found in soils derived from sandstone, limestone, or acid igneous rocks, while a high content is found in soil from glacial sediments or basic igneous rocks. In serpentine soil, concentrations up to several thousand milligrams per kilogram have been reported (Uren 1992). In Denmark, the concentrations in the topsoil mainly range from 1 to 15 mg Ni kg^{-1}, with a mean of approximately 6 mg Ni kg^{-1}, although this depends on the soil texture (Table 2).

Table 2. Soil nickel concentrations (mg Ni kg^{-1} dry wt) in Denmark.

Reference	Soil site (n = number of sites)	Soil concentration (mean) mg kg^{-1}
Friborg (1992)	Natural topsoils (36)	8.8 ± 5.8[a]
Larsen et al. (1996)	Overall (393)	5.0 (0.9 − 15.1)[b]
	Farmland (308)	5.7 (1.3 − 16.2)[b]
	Natural ecosystem soils (14)	1.5 (0.5 − 3.9)[b]
	Deciduous wood (13)	6.2 (1.4 − 13.3)[b]
	Pine wood (30)	1.7 (0.5 − 4.8)[b]
	Mixed wood (19)	5.6 (1.2 − 17.6)[b]
Miljøstyrelsen (1985)	Natural topsoils	6 (0.4 − 15)[c]
Tjell and Hovmand (1978)	Agricultural (44)	6
Århus Amt (1992)	Natural topsoils (71)	(6.7–17)[c]

[a]Mean ± SD (standard deviation).
[b]Mean (5%–95% fractil) (in parentheses).
[c]Mean (range found) (in parentheses).

In Denmark, the soil Ni concentration is highly dependent on the soil texture and especially the clay content (Larsen et al. 1996). Based on 393 samples, including soil Ni content and soil texture, Larsen et al. (1996) derived the following equation for its content in Danish soils:

$$\text{Ni content} = \text{Clay(\%)} \cdot 1.095 - \text{Silt(\%)} \cdot 0.305 - \text{Humus(\%)} \cdot 0.230 + 1.285 \, (r^2 = .88)$$

The dominant factor was the clay content, which showed a correlation with the Ni concentration of $r^2 = .86$. Based on this equation and soil texture data from the Danish Soil Classification (Den Danske Jordklassificering 1976), it was possible to produce a map of the soil Ni concentrations in Denmark (Fig. 1).

V. Sources to the Terrestrial Environment

There are four major sources of Ni to the terrestrial environment: (1) mineralization of soil constituents, (2) atmospheric deposition, (3) distribution of liquids and solids, and (4) decomposition of animal and plant material. Two pathways of Ni removal—leaching to groundwater or surface water and uptake by living organisms—are especially important in agricultural land. No data are available for the extent of Ni leaching from the plowed layer or amounts removed with crops in Denmark. On the national scale in Denmark, aerial deposition (50 tonnes yr^{-1}) has been estimated to account for approximately one-fourth to one-half of the total input (109–207 tonnes yr^{-1}) (Miljøstyrelsen 1985). The most important source of aerial Ni stems from the burning of fossil fuels (Miljøstyrelsen 1985).

On the local scale, the most important sources to the terrestrial environment may be wear of nickel-coated surfaces (13–100 tonnes yr^{-1}), household waste, and loss of coins (21 tonnes yr^{-1}) for nonagricultural soil. Near roads, leaching from coal ash may be an important source if ash is used for road building. For agricultural soils, leaching from commercial fertilizers (8–12 tonnes yr^{-1}), loss from animal food (15 tonnes yr^{-1}), and distribution of sewage sludge (3 tonnes yr^{-1}) are the most important sources (Miljøstyrelsen 1985, 1995). In fields in which sewage sludge is distributed, this is the most important source of nickel (Fig. 2). In an attempt to identify the major sources of Ni in Danish agricultural land from knowledge about its earlier history, i.e., aerial deposition rates, former land usage, and population density, Larsen et al. (1996) were not able to pinpoint any significant sources.

VI. Environmental Fate
A. Profile

Nickel seems, in general, to be uniformly distributed throughout the soil profile. Although this depends on the soil type, it may relate to either the organic matter or the morphic oxides and clay fraction (Kabata-Pendias

Fig. 1. Nickel concentration in Danish soil from an empirical model based on soil concentrations and soil texture (resolution, 1 km^2). Areas enclosed with a thin black line and of uniform pale gray are major towns in which Ni soil concentrations have not been estimated. Shading indicates concentration of Ni in mg kg^{-1} from 0 (light) to >20 (dark).

and Pendias 1984). In forest soils, Ni may accumulate in the top organic layers from decomposition of organic matter. Similar accumulation in top-soil may be seen in sites subjected to heavy aerial Ni deposition. Nickel tends to be less mobile than cadmium but much more mobile than zinc and copper (Tyler and McBride 1982). Its mobility, and thus the fraction dissolved in the soil solution, is highly dependent on the prevailing soil conditions, with pH being very important (Table 3).

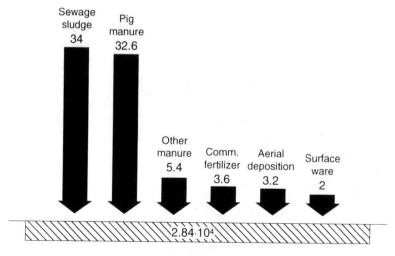

Fig. 2. Deposition of nickel in Danish agricultural soil in g ha^{-1} yr^{-1}. (After Larsen et al. 1996.)

B. Speciation

As with other metals, Ni exists in several different forms in soils and may be bound to exchange sites or specific adsorption sites, adsorbed to or occluded into sesquioxides, fixed with clay mineral lattices, or fixed in organic residues and organisms (Adriano 1986). The binding, and thus the mobility, of Ni in soil depends on several factors, such as the cation-exchange capacity (CEC), pH, texture, organic matter, and $CaCO_3$ content.

Table 3. Mobility of Ni in ten different soils.

Soil	Surface area $(m^2 g^{-1})$	Clay (%)	Fe_2O_3 (%)	Mobility[a]
Clay	67	29	23	H
Silty clay	120	29	5.6	H
Clay	128	40	2.5	H
Clay	51	61	17	H
Sandy loam	122	45	3.7	M
Clay	38	11	1.7	M
Silty clay loam	62	31	4	L
Sand	20	15	1.8	L
Sandy loam	9	5	1.8	L
Loamy sand	8	4	0.6	L

[a]L, Low mobility; M, moderate mobility; H, high mobility.
After Korte et al. (1976).

In general, Ni compounds are relatively soluble at pH below 6.5; at pH above 6.7, Ni exists as insoluble nickel hydroxides that are slowly mobile (Schmitt and Sticher 1992). Acid rain, therefore, has a pronounced tendency to mobilize Ni from soils and causes increased levels in groundwater (Sunderman and Oskarsson 1991). The divalent ion, Ni(II), seems to be the most stable form in soil solutions (Uren 1992). In soil solution, it may occur in the ionic form or complexes with either organic or inorganic ligands. In acidic soil, the most dominant soil solution form seems to be Ni^{2+} or $NiSO_4$, whereas in alluvial soils $NiCO_3$, $NiHCO_3$, and Ni^{2+} seem to be the most stable forms (Uren 1992). In Canadian agricultural soils, Whitby et al. (1978) found 0.5%-2% of the total Ni to be extractable with diethylenetriamine penaacetic acid (DPTA). Higher results, 2%-5% extracted by DTPA, were found by Sadiq (1985), who spiked soil with Ni and left it for 6 mon before extraction. When extracted with water, less than 0.5% is recovered (Haq et al. 1980). Uren (1992) reported that approximately 0.001% of the total Ni in soil is in solution.

VII. Bioaccumulation, Bioconcentration, and Biomagnification

The potential for a soil-living organism to bioaccumulate or bioconcentrate, that is, to accumulate compounds to concentrations higher than in the ambient environment, is highly dependent on the species involved and the environment in which it lives. Some general conclusions can be made, however.

A. Plants

It is well established in the literature that plants accumulate Ni in response to its availability in the soil, but quantitatively the soil–plant relationships depend on the species involved. In plants from uncontaminated soil, the aboveground tissue concentrations have usually been found to be less than 5-10 mg Ni kg^{-1} (Adriano 1986; Lepp 1980; McIlveen and Negusanti 1994). Some plants can bioaccumulate Ni to higher levels than are present in the soil, but for most plants the bioaccumulation factor is less than 1.0. Sauerbeck and Styperek (1988; as cited by Grün et al. 1994) found aboveground concentration factors to be less than 1.0 for many crops, except for carrots and kale, which had bioaccumulation factors between 1.0 and 2.0. Underground tissue may, however, accumulate high concentrations of Ni and have bioaccumulation factors greater than 1.0. Some species, termed hyperaccumulators, may bioaccumulate extremely high concentrations (more than 1000 mg Ni kg^{-1}) in the aboveground tissue. Hyperaccumulators are mainly found where ultrabasic substrates are involved, such as serpentine soils (see Boyd and Martens 1994; Brooks 1980; IPCS 1992; Lepp 1980; McIlveen and Negusanti 1994). Recent investigations indicate that free histidine in the xylem sap is responsible for the tolerance of hyperaccumulators to Ni (Krämer et al. 1996). It is unknown what function such hyperaccumulation of Ni has for the plant, but one theory is that Ni constitutes a defense mechanism against herbivory. Supporting this, Boyd and Martens (1994) showed that Ni bioaccumulated by *Thlaspi montanum* was acutely toxic to the cabbageworm *Pieris rapae*.

B. Invertebrates

The bioaccumulation of Ni by soil-dwelling invertebrates has been described in several studies, but mainly earthworm species have been investigated. Nickel, in general, does not seem to bioaccumulate in terrestrial invertebrates and has bioaccumulation factors well below 1.0, although exceptions are known (Hopkin 1989; Neuhauser et al. 1995). In two studies applying Ni to sewage sludge, added as a layer on top of the soil, Hartenstein et al. (1980) and Neuhauser et al. (1983) found bioaccumulation factors much lower than 1.0, although concentrations in the animals increased with increasing ambient concentrations. In both studies, the worms may have avoided the contaminated top layer, which may have reduced exposure. Ma (1982) investigated the influence of soil properties and worm-related factors on the bioaccumulation of heavy metals in the earthworm *Lumbricus rubellus* and found that the bioaccumulation factor almost never exceeded 1.0 and was negatively correlated with soil pH and CEC. Also investigating a *Lumbricus* species, Beyer et al. (1982) observed that concentrations of Ni were less than or equal to the ambient concentration. In contrast to this, Gish and Christensen (1973) investigated soil and earthworm levels adjacent to three different roads. They found soil concentrations ranging from 12 to 26 mg Ni kg^{-1} and worm concentrations between 12 to 38 mg Ni kg^{-1}, with an average bioaccumulation factor of 2.0. Investigating the uptake and excretion of metals in worms, Neuhauser et al. (1995) found that increasing soil Ni concentrations led to increasing internal concentrations in earthworms but that Ni was eliminated rather rapidly if the worms were transferred to clean soil. Although most studies show little accumulation of Ni in invertebrates, some species exhibit large internal concentrations. Studying a termite mound in a serpentine soil, Wild (1975) reported concentrations of up to 1500 mg Ni kg^{-1} in termites and to 7700 mg Ni kg^{-1} in a tenebrionid beetle. It would be interesting to see whether free histidine is responsible for binding Ni in the tissue fluid of such high-accumulating invertebrates, thus leading to increased tolerance, as is the case for hyperaccumulating plants. It may be speculated, as for hyperaccumulating plants, that organisms with high internal Ni concentrations, such as the tenebrionid beetle, may utilize the high internal Ni level as a defense against predators.

C. Biomagnification

It is not possible at present to verify whether Ni is biomagnified. However, the low bioaccumulation factors strongly indicate that biomagnification does not present a problem for Ni in the lower terrestrial food chain.

Nevertheless, as previously pointed out by Janssen et al. (1993), there is a need to study specific food chains in order to verify the possibility of biomagnification.

VIII. Ecotoxicological Effects
A. General Aspects

Nickel is considered to be essential to animals, microorganisms, and plants. Concentration levels required to maintain normal function are not known, but Ni has been shown to be a constituent of enzymes and proteins, e.g., urease methyl coenzyme reductase M, hydrogenase, and carbon monoxide dehydrogenase, in microorganisms (Hausinger 1992; Welch 1995).

Because Ni is an essential element, either deficiency or toxicity symptoms can occur when too little or too much Ni, respectively, is taken up (Hopkin 1989). The toxic functions of nickel probably result primarily from its ability to replace other metal ions in enzymes and proteins or to bind to cellular compounds containing O-, S-, and N-atoms, such as enzymes and nucleic acids, which are then inhibited. According to the definition by Nierboer and Richardson (1980), Ni belongs, as do chromium and zinc, to the group of "borderline metals," which indicates that Ni may replace "class A" and other borderline metals. Such substitution may cause malfunctions of both metal-containing and metal-activated enzymes.

The effect levels, and to some extent the toxic modes, are dependent on many factors, such as the physiological condition of the organism exposed and the physiochemical characteristics of the ambient environment. Concentrations that cause deficiency or toxic symptoms may also depend on the presence of other metals and their interactions. Interactions with other metals are known to be prominent features for all metals. Cataldo et al. (1978) found that Co, Cu, Fe, and Zn influenced the adsorption of Ni to soybeans.

B. Microorganisms and Microbial Processes

Studies dealing with the effect of Ni on microorganisms and microbial processes may be divided into four groups: (1) those concerned with carbon mineralization (C mineralization); (2) those concerned with nitrogen mineralization (N mineralization) and nitrification; (3) those about enzymatic processes; and (4) those concerned with microbial numbers or biomass. A summary of studies involving microorganisms is shown in Table 4.

Carbon Mineralization. C mineralization, or respiration, in soils is known to be affected by a wealth of pollutants, of which Ni is no exception. In soil devoid of plant roots, respiration is almost completely dependent on the activity of soil microbes, mainly bacteria and fungi (Tyler et al. 1992). Soil respiration is performed by many microorganisms and thus is a nonspecific measure of soil microbial activity.

Table 4. NOEC, LOEC, and EC_x values for effects of Ni on microorganism and microbial processes in soil.

Reference	Soil	Compound	Time	Parameter	Effect data (mg Ni kg^{-1} dry wt soil)	
					NOEC	LOEC/EC$_{(x)}$
Al-Khafaji and Tabatabi (1979)	Nicollette Webster Harps	NiCl$_2$	1 hr	Enzyme activity	— — 1470	1470$_{(26)}$ 147$_{(3)}$
Babich and Stotzky (1982)	— Montmorillonite Kaolinite —	NiCl$_2$	1–8 d	Microbial (fungi) numbers	10 (30 T) 1000 (1017 T)	50 (80 T) 250 (280 T) 750 (780 T) 750 (780 T)
Beck (1981)	—	NiSO$_4$	3 d 30 d 7 d 38 d	Microbial numbers Nitrification	50 50	50$_{(15-28)}$ 150$_{(14-59)}$ 50$_{(10)}$ 150$_{(19)}$
Bhuiya and Cornfield (1972)	Bagshot sand	NiSO$_4$	12 wk	C mineralization		1000
Bremner and Douglas (1971)	Silty clay loam	NiCl$_2$	5 hr	Enzyme activity	50$_{(1-2)}$ N	
Chander and Brookes (1991)	Sandy loam Silty loam	Ni sludge	22 yr	Microbial numbers	164 T 258 T	
Chaudri et al. (1992)	Sandy loam	NiSO$_4$	2 mon	Microbial numbers	30 (54 T)	
Cornfield (1977)	Loamy sand	NiSO$_4$	2 wk 8 wk	C mineralization		10$_{(18)}$ 10$_{(12)}$

Reference	Soil	Compound	Time	Parameter		
De Catanzaro and Hutchinson (1985)	—	$NiSO_4$	2 wk	Microbial numbers / N mineralization / Nitrification	500 / 500 / 100	500
Doelman and Haanstra (1984)	Sandy loam	$NiCl_2$	90 wk	C mineralization	400 (402 T)	1000 (1002 T)
Doelman and Haanstra (1986)	Sand	$NiCl_2$	6 wk/18 mon	Enzyme activity		$30_{(10)}/120_{(10)}$ N
	Sandy loam		6 wk/18 mon			$860_{(10)}/2300_{(10)}$ E,N
	Silty loam		6 wk/18 mon			$130_{(10)}/14_{(10)}$ E,N
	Clay		6 wk/18 mon			$610_{(10)}/90_{(10)}$ E,N
	Sandy peat		6 wk/18 mon			$1100_{(10)}/540_{(10)}$ E,N
El-Sharouny et al. (1988)	Clay	$NiCl_2$	15 wk	Microbial number (*, increase observed)	5000*	500
Frankenberger and Tabatabai (1981)	—	$NiCl_2$	0.5 hr	Enzyme activity	29	$294_{(6)}$
Frankenberger and Tabatabai (1991)	—	$NiCl_2$	0.5 hr	Enzyme activity		$294_{(18-23)}$
Fu and Tabatabai (1989)	—	$NiCl_2$	24 hr	Enzyme activity	147	$147_{(15-63)}$
Frostegård et al. (1993)	Sandy			Enzyme activity		$881_{(50)}$
						$94_{(10)}$
				C mineralization		$353_{(50)}$
						$88_{(10)}$
				Microbial number		$88_{(10)}$

(continued)

Table 4. (*Continued*)

Reference	Soil	Compound	Time	Parameter	Effect data (mg Ni kg^{-1} dry wt soil)	
					NOEC	LOEC/EC$_{(x)}$
Giashuddin and Cornfield (1978)	Sandy	NiSO$_4$	6 wk	C-/N mineralization Nitrification		10 10
Giashuddin and Cornfield (1979)	Sandy	NiO	6 wk	C-/N mineralization		50
Gupta et al. (1987)	Sandy loam	NiSO$_4$	1–63 d	C mineralization	27	90$_{(20)}$
Haanstra and Doelman (1984)	Sand	NiCl$_2$	18 mon	Enzyme activity	55 (63 T)	400 (408 T)
	Silty loam		18 mon			55 (80 T)
	Clay		18 mon		55 (93 T)	400 (439 T)
	Sandy peat		18 mon		55 (59 T)	400 (404 T)
Haanstra & Doelman (1991)	Sand	NiCl$_2$	6 wk/18 mon	Enzyme activity		372$_{(10)}$/55$_{(>10)}$ N
	Sandy loam		6 wk/18 mon			611$_{(10)}$ E,N
	Silty loam		6 wk/18 mon			2207$_{(10)}$/55$_{(>10)}$ N
	Clay		6 wk/18 mon			1069$_{(10)}$/271$_{(10)}$ E,N
	Sandy peat		6 wk/18 mon			7045$_{(10)}$ E,N
Juma and Tabatabi (1977)	—	NiCl$_2$	0.5 hr	Enzyme activity		147$_{(8-11)}$

Reference	Soil	Salt	Duration	Process		
Liang and Tabatabai (1977)	—	$NiCl_2$	20 d	N mineralization		$294_{(7-17)}$
Liang and Tabatabai (1978)	—	$NiCl_2$	10 d	Nitrification		$294_{(62-67)}$
Lighthart et al. (1983)	—	$NiSO_4$	—	C mineralization	29 N	294 N
Rogers and Li (1985)	—	$NiSO_4$	28 d	Enzyme activity		$77_{(50)}$ E $30_{(39)}$ $114_{(50)}$ E $30_{(14)}$
	—	I				
Spalding (1979)	Douglas-fir needle litter	$NiCl_2$	4 wk	C mineralization	100	1000
				Enzyme activity	1000	
Stadelmann & Santchi-Fuhrimann (1987)	Sandy loam	$NiCl_2$	62 d	C mineralization	17 N (31 T),N	$56_{(20)}$ N (70T)
Stott et al. (1985)	—	$NiCl_2$	5 hr	Enzyme activity	1468	
Tabatabi (1977)	Okoboji Harps	$NiCl_2$	2 hr	Enzyme activity	29 N	$29_{(13)}$ N $290_{(20-33)}$ N
Wilke (1988)	Sandy cambisol	$NiCl_2$	9 yr	Enzyme activity	100	100 400
Yadav et al. (1986)	Clay loam	Ni	12 hr	Enzyme activity		$100_{(13)}$

NOEC, No-observed-effect concentration; LOEC, lowest-observed-effect concentration; EC_x, concentration at which x% effect is observed. hr, Hours; d, days; wk, weeks; mon, months; yr, years; T, total concentrations; E, extrapolated value; N, no significance stated.

As is seen in the following, the toxic levels for Ni range over a factor of 100, from approximately 10 to more than 1000 mg Ni kg^{-1}, probably depending, among other factors, on the microbial community, soil type, and exposure time. Investigating C mineralization in a sandy soil 6 wk after Ni was added as $NiSO_4$, Giashuddin and Cornfield (1978) observed a 22% reduction of C mineralization at 10 mg Ni kg^{-1}, the lowest level of addition, which suggests an even lower no-effect level. Using NiO salt, rather than $NiSO_4$ salt, Giashuddin and Cornfield (1979) observed a reduced C mineralization only at additions of 50 mg Ni kg^{-1} or greater. In an extensive study of the effects of various metals on C mineralization in a sandy loam, Stadelmann and Santschi-Fuhrimann (1987) found a no-effect level at additions of 13 mg Ni kg^{-1} (making the total content 30 mg Ni kg^{-1}) and a 20% reduction at addition of 56 mg Ni kg^{-1} (total, 70 mg Ni kg^{-1}). In another experiment on the C mineralization of a sandy loam, Gupta et al. (1987) calculated a no-effect level of 27 mg Ni kg^{-1}, based on exposure intervals from 1 to 63 d.

Similar observations were made by Lighthart et al. (1983), who found an effect threshold between 29 and 293 mg Ni kg^{-1} (0.5-5 mmol kg^{-1}). Studying the carbon dioxide release from a sandy soil under incubation with 12 different metals added individually at two time intervals, Cornfield (1977) observed a reduction in CO_2 release when 10 mg Ni kg^{-1} was added (as $NiSO_4$) after both 2 and 8 wk. The reduction was 18% after 2 wk but only 6% reduction was observed after 8 wk. The reduction in toxicity with time may have been because some types of heterotrophs could adapt to some extent late in incubation or because of reduced availability (Cornfield 1977). On addition of 100 mg Ni kg^{-1}, the effect increased with time from 12% to 28% reduction, which suggested a slow binding of Ni to the soil at this level.

Also investigating the effects at different intervals after addition of $NiCl_2$, Doelman and Haanstra (1984) found no-observable-effect concentrations (NOEC) ranging from 400 to 1000 mg Ni kg^{-1}. In this study, the effect concentrations started at 1000 mg Ni kg^{-1} and above. Clay, sandy loam, and sandy peat soils exhibited more pronounced effects on a short-term basis compared to long-term exposures. Three other studies reported effects only at higher levels; Spalding (1979) observed a reduction of the CO_2 production in Douglas-fir needle litter following addition of 1000 mg Ni kg^{-1}, while no effect was observed following addition of 100 mg Ni kg^{-1}. Bhuiya and Cornfield (1972) found a reduction in the release of CO_2 2 mon after 1000 mg Ni kg^{-1} was added to a sandy soil. Hattori (1992) mixed 587 mg Ni kg^{-1} into soil and added sewage sludge after 3 d, which resulted in a reduction in the CO_2 release as observed by repeated measurements during 4 wk.

Nitrogen Mineralization and Nitrification. The N mineralization, or the conversion of organic N compounds, in soil is accomplished by most heterotrophic soil organisms as for C mineralization, whereas nitrification (the oxidation of NH_4^+ to NO_2^- and NO_3^-), is mainly carried out by a few specialized bacteria (Tyler et al. 1992).

Giashuddin and Cornfield (1978) investigated N mineralization and nitrification in a sandy soil for 6 wk after addition of Ni and observed a 22% reduction of N mineralization at the lowest level of addition, 10 mg Ni kg^{-1}. This result suggests a no-effect level lower than 10 mg kg^{-1}. With increasing Ni levels, the nitrification decreased to a greater extent than did N and C mineralization, although the toxic effects were little different between 100 and 1000 mg Ni kg^{-1}. Using NiO rather than $NiSO_4$, Giashuddin and Cornfield (1979) observed reduced N mineralization at addition of 50 mg Ni kg^{-1}. Beck (1981) investigated the nitrification within 7 d after addition of Ni and observed a 10% reduction after 50 mg Ni kg^{-1} and a 20% reduction following addition of 150 mg Ni kg^{-1}. This reduction was stable for as long as 38 d.

Liang and Tabatabai (1977) studied the effects of several metals on N mineralization in U. S. soils. At 20 d after adding about 294 mg Ni kg^{-1} (5 μmol g^{-1}), N mineralization was reduced from 7% to 17%, depending on the soil type.

At concentrations and incubation times similar to those of the previous experiment, Liang and Tabatabai (1978) evaluated the effect of trace metals on nitrification in these soils and observed a 62%–67% inhibition. Thus, nitrification seems more sensitive than N mineralization. De Catanzaro and Hutchinson (1985) studied the effect of Ni on nitrification and N mineralization in soils from three boreal jack pine forests. The potential rates of nitrification and N mineralization were extremely low in these soils, and, in contrast to the studies cited above, these authors found that 100 mg Ni kg^{-1} could stimulate nitrification, with an inhibition at 500 mg Ni kg^{-1}, and that addition of 500 mg Ni kg^{-1} could stimulate N mineralization.

Enzymatic Processes.   A variety of enzymatic activities have been measured in soil. Such enzymatic processes usually originate from unknown sources but are thought mainly to be extracellular products from microorganisms and plants (Tyler et al. 1992). Enzymatic processes are responsible for a wealth of activity in soil; e.g., acid and alkaline phosphatase are thought to be responsible for the recirculation of phosphorus (P) in the ecosystem. Studies considering the toxic effects of Ni on enzymatic processes may be divided into several subgroups: nitrate reductase activity (NRA), acid and alkaline phosphatase activity (APA), urease activity (UA), L-asparaginase activity (LAA), adenosine triphosphate (ATP) content, dehydrogenase activity (DHA), saccharase activity (SA), protease activity (PA), arylsulfatase activity (ASA), and glutamic acid decomposition rates (GAD).

The activity of acid and alkaline phosphatase, in catalyzing the hydrolysis of organic P compounds is of great importance to the recirculation of P in all ecosystems. Juma and Tabatabai (1977) studied the effect of 1468 mg Ni kg^{-1} (25 μmol g^{-1}) and 147 mg Ni kg^{-1} (2.5 μmol g^{-1}), applied as NiCl$_2$, on phosphatase activity in three different U.S. soils. At 25 μmol/g, reduction was about 20%, and about 10% with 147 mg Ni kg^{-1}. At the higher concentration, no differences in sensitivity were found between the different soils for total acid phosphatase activity, but a twofold difference was found in the alkaline phosphatase activity of soils with high organic matter content (5.45%) and soils with a lower organic matter content (3.74%), with alkaline phosphatase activity being most sensitive in the latter. In a long-term study (9 yr), Wilke (1988) found significant reductions, as much as 48%, in the alkaline phosphatase activity when applying 100 and 400 mg Ni kg^{-1} to a sandy soil with low organic carbon content. Stott et al. (1985), in studying the effect of nickel (NiCl$_2$) and other metals on the pyrophosphatase activity in three U.S. soils, observed no significant reductions at concentrations of 296 mg Ni kg^{-1} (5 μmol g^{-1}) or 1468 mg Ni kg^{-1} (25 μmol g^{-1}).

Urea is added to soils as a synthetic fertilizer and is a constituent of animal excreta. Inhibition of the decomposition process, the urease activity, has been studied in several experiments. This enzyme usually originates from plant sources, and many metal ions have been shown to be potent inhibitors of its activity. Bremner and Douglas (1971) investigated the effect of different metallic and organic compounds on urease activity in three U.S.soils; following addition of 50 mg Ni kg^{-1}, only a 1%–2% inhibition was noticed. Tabatabai (1977) also studied the effect of several metals on urease activity in different U.S. soils and found that upon addition of 296 mg Ni kg^{-1} (5 μmol g^{-1}) reduced urease activity 20%–33%, with the greatest reduction occurring in a soil of pH 5.1. No apparent trend between soil composition and degree of inhibition was noticed, although urease activity itself was correlated with organic carbon content. A 13% reduction in urease activity was observed when 30 mg Ni kg^{-1} (0.5 μmol g^{-1}) was added to a soil with high organic carbon content.

Doelman and Haanstra (1986) investigated five different Dutch soils and calculated a 6-wk EC$_{10}$ ranging from 30 to 1100 mg Ni kg^{-1}, depending on soil type. After 1.5 yr, the EC$_{10}$ values ranged from 14 to 2300 mg Ni kg^{-1}, but the relative sensitivity differed from that of the 6-wk study. In the former test, a sandy peat soil was the least sensitive, whereas in the latter a sandy loam was the least sensitive (see Table 4). Likewise, the soil with the most sensitive processes in the former test was the sandy soil, whereas in the latter it was a silty loam. For two of the five soils, the sensitivities were greater after 6 wk than after 1.5 yr.

A number of enzymatic processes have been investigated very little and often are presented only in one or two papers. The enzyme amidase, which catalyzes the hydrolysis of aliphatic amides producing NH$_3$ and CO$_2$, and its substrates, may be a potential fertilizer of soils. For this enzyme, Frankenberger and Tabatabai (1981) observed no significant effects at 29.3 mg

Ni kg^{-1} (0.5 μmol g^{-1}), whereas a 6% reduction in activity was observed when adding 293 mg Ni kg^{-1} (5 μmol g^{-1}), although only in the most organic soil. Nitrate reductase is an important enzyme in the process of denitrification, and inhibition of this enzyme may reduce denitrification. Studying this in a range of U.S. soils, Fu and Tabatabai (1989) found that the greatest effect on activity in sandy soils was a 63% reduction at 147 mg Ni kg^{-1} (2.5 μmol g^{-1}), while the effect was less pronounced (15% reduction) in a soil with less sand and higher organic content. No effect was found in an organic soil with a pH of 7.8.

Blagoveshchenskaya and Danchenko (1974; cited by Frankenberger and Tabatabi 1991) found that L-asparaginase catalyzes the hydrolysis of L-asparagine with the production of ammonia and aspartic acid and plays an important role in N mineralization. Adding 293 mg Ni kg^{-1} (5 μmol Ni g^{-1}), reductions ranged from 18% to 23% in different soils (Frankenberger and Tabatabai 1991). The L-asparaginase activity correlated with organic C and total N but not with clay or sand. Wilke (1988) investigated a range of microbial endpoints, one of these being the determination of biomass by soil ATP content. Using a sandy cambisol, no inhibition of ATP activity at 400 mg Ni kg^{-1} was found. The soil dehydrogenase activity (DHA), which represents an oxidoreductase enzyme that catalyzes the oxidation of substrates, thus producing electrons that can enter the cell electron transport system, may also be used as a measure of microbial activity. Rogers and Li (1985) found a significant reduction of DHA at 30 mg Ni kg^{-1} and an EC$_{50}$ of 77 mg Ni kg^{-1}. Upon enrichment of the soil with 1% alfalfa, the effect was less and the EC$_{50}$ was 144 mg Ni kg^{-1}. In comparison, Wilke (1988) observed a 37% reduction in DHA after adding 100 mg Ni kg^{-1} to a sandy cambisol.

Arylsulfatases are enzymes that catalyze the hydrolysis of arylsulfate anions by fission of O–S bonds and are thus important for S-cycling in soils (Al-khafaji and Tabatabai 1979). Haanstra and Doelman (1991) studied the effect of Ni, among other metals, on the arylsulfatase activity in different soils after 6 wk and 1.5 yr of exposure. Adding from 55 to 8000 mg Ni kg^{-1}, using seven concentrations, they found EC$_{10}$ values from 372 mg Ni kg^{-1} (6.34 mmol kg^{-1}) to 2207 mg Ni kg^{-1} (37.6 mmol kg^{-1}) after 6 wk, and 1.2 mg Ni kg^{-1} (0.02 mmol kg^{-1}) to 7045 mg Ni kg^{-1} (120 mmol kg^{-1}) after 1.5 yr. Activity in sand was the most sensitive, whereas in clay and especially in sandy peat this process was much less sensitive. In some cases, no effect could be detected at any concentration. Al-khafaji and Tabatabai (1979) observed a 3% reduction with 147 mg Ni kg^{-1} after a 30-min exposure, while adding 1470 mg Ni kg^{-1} resulted in a 26% reduction of arylsulfatase activity, the magnitude depending on soil type.

Haanstra and Doelman (1984) reported the toxicity of several metals on the decomposition of soil glutamic acid (an amino acid). For Ni, an increased decomposition time was present at addition of 55 mg Ni kg^{-1} after 6 mon with increasing (linear) decomposition time up to 2000 mg Ni kg^{-1}, while after 1.5 yr, effects were only significant at or above 400 mg Ni kg^{-1}. Wilke (1988) investigated saccharase activity and found a reduction only with 400 mg Ni kg^{-1}. Protease seems to be less sensitive, as Stott et al. (1985) observed no effect of Ni upon protease activity, even at an application of 1470 mg Ni kg^{-1} (25 μmol g^{-1}), and Wilke (1988) found no effect at 400 mg Ni kg^{-1}. When Spalding (1979) measured the effect of Ni salts on cellulase, xylanase, amylase, invertase, β-glucosidase, and polyphenoloxidase activity in Douglas-fir needle litter, only the polyphenoloxidase activity was affected significantly by Ni, and this only temporarily.

Microbial Numbers. As seen from the preceding section, microbial processes are sensitive to Ni. Such toxic effects may also be expressed as a reduction in the number of microbes or a shift in community composition. A shift in microbial composition from both gram-negative and gram-positive organisms to a community dominated by gram-negative organisms may often be observed upon pollution (Frostegård et al. 1993).

Frostegård et al. (1993) used a change in the phospholipid fatty acid (PLFA) patterns as an indicator of shifts in the microbial community. They found a change in the microbial community of a sandy soil and a forest soil upon addition of 88 mg Ni kg^{-1}, which was comparable to the sensitivity of the respiration and the ATP level. By contaminating soil with Ni for 3 d, Beck (1981) found 15%–30% reduction in the microbial biomass in soils when tested at 50 mg Ni kg^{-1} and 30%–60% reduction at 150 mg Ni kg^{-1}; these effects were less severe after 30 d. For growth of filamentous fungi in a Kitchawan soil, Babich and Stozsky (1982) found a reduction in the microbial population at 50 mg Ni kg^{-1}. In comparison, growth and survival of other microorganisms were reduced at 250 mg Ni kg^{-1} (the lowest concentration tested). In one soil, no effect on microbial growth was observed even at 1000 mg Ni kg^{-1}. Addition of montmorillonite or kaolinite (clay), which increased the binding capacity of Ni, reduced the effect at 250 mg Ni kg^{-1}. Chaudri et al. (1992) investigated survival of the indigenous population of *Rhizobium leguminosarum* in soil spiked with Ni salts and found Ni had no appreciable effect on the number of rhizobia at 60 mg Ni kg^{-1}. Chaudri et al. (1993) investigated the effect on the indigenous population in soil treated with metal-contaminated sewage sludge. Effects were observed, but it was not clear which chemical compound caused the effects, although it was not likely Ni. Also, using sludge contaminated with Ni, Chander and Brookes (1991) measured microbial biomass carbon 22 yr after addition to two different soils and found no effects when Ni concentrations were increased three- to fourfold, i.e., from 34 to 162 mg Ni kg^{-1} and from 53 to 258 mg Ni kg^{-1}, although other metals were also present in increased concentrations. In contrast to the foregoing observations, a few studies described an increase in microbial populations with addition of Ni. De Catanzaro and Hutchinson (1985) estimated the microbial population by the most probable number (MPN) method and found that the nitrifier population, which was extremely low in uncontaminated soils, increased after adding Ni. El-Sharouny et al. (1988) found an increase in the cell count of fungi in soil at 5000 mg Ni kg^{-1}.

C. Plants

The toxicity of excessive Ni toward plants was first pointed out by Haselhof in 1893, who demonstrated its toxicity to corn and bean plants using a solution culture technique (Lepp 1980). It has since been found that Ni commonly induced patchy discoloration, premature senescence, and yellowing of old foliage, and several authors have found that Ni affects the Fe and P status of plants (Adriano 1986; Mishra and Kar 1974; Uren 1992). Most of such effects have been investigated in solution cultures, which have little resemblance to conditions for plants in the soil environment. Therefore, only investigations of growth in a soil medium are discussed here. A summary is given in Table 5.

When reviewing the literature, it is noticed that most toxicity tests are concerned mainly with yield effects, and few other toxicological parameters are used. It is desirable to have toxicity data for other and perhaps more sensitive endpoints, such as quantity of seeds produced, metabolic rate, seed germination, and biochemical endpoints. It is also noticed that few wild species have been tested (Table 5).

Oat, Wheat, and Barley.   Using nickel chloride, Halsted et al. (1969) investigated the effect of Ni on oats. Nickel chloride was added at 0, 20, 50, 100, and 500 mg Ni kg^{-1} to different soils and left for 7 mon for incubation, with addition of lime and fertilizers after 1 mon. Effects were present at 100 mg Ni kg^{-1}, but only for plants grown in upland sand and not those grown in sandy loam. Addition of lime reduced the toxicity of Ni. Webber (1972) investigated the effects of Ni on oats in a silty loam; increasing concentrations of Ni reduced fresh weight yield, more so at pH 5.7 than at pH 6.4. At 50 mg Ni kg^{-1}, a decrease in yield of about 15% was observed. Mean height was reduced at 37.5 mg Ni kg^{-1} when Ni was mixed into sewage sludge.

Using three different soil types, Sorteberg et al. (1974) found no observable effects on yield at the lowest concentration (50 mg Ni kg^{-1}) and no difference in toxicity levels among the three soil types used in this experiment. Using nickel acetate, de Haan et al. (1985) observed differences in the toxic effect level of Ni on oats in several different soils. They found significantly reduced yield at 12.5 mg Ni kg^{-1} in a soil with a low clay content, although conflicting statements are found in their report. The corresponding no-effect level was 6.25 mg Ni kg^{-1}. In soils with higher clay or organic matter content effects were not observed until at least 25 mg Ni kg^{-1} was added. Plants grown in a soil containing about 20% organic matter did not show any yield decrease up to 100 mg Ni kg^{-1}. Using higher

Table 5. NOEC, LOEC, and EC_x values for effects of Ni on plants in soil.

Reference	Soil	Compound	Time	Parameter	Species	Effect data (mg Ni kg^{-1} dry wt soil) NOEC	LOEC/ $EC_{(x)}$
Allison and Dzialo (1981)	Sandy loam	Ni(NO$_3$)$_2$	2 mon	Yield	Oat (Avena sativa)	50	
Bingham et al. (1979)	Sandy loam, limed	NiSO$_4$	90	Yield	Wheat (Triticum aestivum)	80	
Dahiya et al. (1994)	Sandy	NiCl$_2$	10 wk	Yield	Wheat (Triticum aestivum) (* compared to 10)	10 N	20 N*
				N content		7,5 N	10 N
Dang et al. (1990)	Clay loam	NiSO$_4$	8 wk	Yield	Onion (Allicum cepa)		50$_{(20)}$
					Fenugreek (Trigonella poenum gracum)		50$_{(20)}$
de Haan et al. (1985)	– I/II/III/V Ni – IV (C$_2$H$_5$O$_2$)$_2$ – VI	6 mon	Yield	Oat	25 6.25 100	50 12.5
Gupta et al. (1987)	Sandy loam Sandy loam	NiSO$_4$	–	Yield	Lettuce (Lactuca sativa)	29 172–335	46 253–503

Frossard et al. (1989)	Sandy loam	$NiSO_4$	4 wk	Yield/carbohyrate content	Ryegrass (*Lolium perenne*)	50 (69 T)	100 (119 T)
Halstead et al. (1969)	– I	$NiCl_2$	110 d	Yield	Alfalfa	20 N	50 N
	– II				Oat	50 N	100 N
					Alfalfa/oat	100 N	500 N
Khalid and Tinsley (1980)	Loam	$NiSO_4$	4 wk	Yield	Ryegrass (*Lolium perenne*)	30 N	$90_{(14)}$ N
MacLean and Dekker (1978)	Loam	$NiSO_4$	6 wk	Yield	Corn		60
	Clay					60	240
					Lettuce (*Lactuca sativa*)	240 N	30 N
							480 N
Metwally and Rabie (1989)	Alluvial	$NiCl_2$	50 d	Yield	*Fabia* bean	50 (104 T)	160 (224 T)
					Corn	120 (174 T)	
	Sandy loam				*Fabia* bean	50 (90 T)	
					Corn	40 (80 T)	80 (120 T)
Mitchell et al. (1978)	Sandy loam	$NiSO_4$	8 wk	Yield	Lettuce (*Lactuca sativa*)		$60_{(25)}$
			15 wk		Wheat (*Triticum aestivum*)		$102_{(25)}$
Webber (1972)	Silty loam	$NiSO_4$	–	Yield	Oat/mustard		$50_{(15/30)}$ N
Patterson and Olson (1983)	Mineral	$NiSO_4$/ $NiNO_3$	5–21 d	Radical growth	White spruce (*Picea marina*)		50
					White pine (*Pinus strobulus*)	50	100

(continued)

Table 5. (*Continued*)

Reference	Soil	Compound	Time	Parameter	Species	Effect data (mg Ni kg^{-1} dry wt soil)	
						NOEC	LOEC/ EC$_{(x)}$
Roth et al. (1971)	Peaty muck	NiSO$_4$	31 d	Yield	Oat (*Avena sativa*)	3757	7515
			46 d		Soybeans (*Glycine max*)	939	1879
Singh and Jeng (1993)	Sandy	NiCl$_2$	3–12 wk (3 yr)	Yield	Ryegrass	50	
Sorteberg (1974)	Clay Sphagnum Sandy	NiCl$_2$	—	Yield	Oat	50 N 50 N 50 N	250 N 250 N 250 N
Taylor and Allison (1981)	Sandy loam	Ni(NO$_3$)$_2$	100 d	Yield	Alfalfa (*Medicago sativa*)	50 N	125 N
Wallace et al. (1980)	Loam	NiSO$_4$	20 d	Yield	Corn (*Zea mays*) Barley (*Hordeum vulgare*) Soybeans (*Glycine max*)	100 100 100	

concentrations, Roth et al. (1971) investigated the toxic effect and accumulation of Cu and Ni in oats grown in a peat soil. Nickel only reduced the oat dry weight yield after 7515 mg Ni kg^{-1} was added, with no effects at 3757 mg.

Investigating the effect of Ni on wheat, Mitchell et al. (1978) found that about 300 mg Ni kg^{-1} caused 25% reduction for plants grown in a silt loam, but only 102 mg Ni kg^{-1} was needed in an acid fine sandy loam. Bingham et al. (1979) observed reduced wheat yield ($\sim$ 10%–30%) at 80 mg Ni kg^{-1} in unlimed soils. In this experiment, 25 mg Ni kg^{-1}, Cu, and Zn were also added. In limed soils, however, similar metal additions did not cause alterations in yield. With increasing concentrations of Ni, up to 7.5 mg Ni kg^{-1} in soil, Dahiya et al. (1994) observed increased N uptake by wheat following addition of N. At higher Ni concentrations, 10 mg and 20 mg Ni kg^{-1}, N uptake decreased. Yield was reduced by addition of 20 mg Ni kg^{-1}, regardless of the N addition, when compared to the yield at 10 mg Ni kg^{-1} but not when compared to the control. Wallace et al. (1980) found a yield reduction in barley following addition of 100 mg Ni kg^{-1}; a significant synergistic effect on barley yield was observed when six metals were added at the same time.

Ryegrass. Two studies showed no effects on the yield of ryegrass following addition of 50 mg Ni kg^{-1}. Allison and Dzialo (1981) estimated the effect on ryegrass after 2, 5, and 7 mon growth in a sandy loam and found no effects on yield. Likewise, Singh and Jeng (1993) found no effects on ryegrass yield when grown in a sandy soil with addition of Ni salts to 50 mg Ni kg^{-1}. Nickel was added only the first year, and plants were grown for 3 successive years; accumulation decreased with time. When growing ryegrass for 4 wk in a loam soil, no effects were found on yield at 30 mg Ni kg^{-1}, but a depression of shoot yields at 90 mg Ni kg^{-1} and more was seen (Khalid and Tinsley 1980). Frossard et al. (1989) studied the effects of different heavy metals on fructan, sugar, and starch content in ryegrass and showed that the carbohydrate components were affected at the same level as the yield, 100 mg Ni kg^{-1}, but although fructan content increased by 25%, sucrose content decreased by 38%.

Lettuce. Mitchell et al. (1978) observed a 25% reduction in lettuce yield at approximately 300 mg Ni kg^{-1} in a silt loam, whereas only 60 mg was needed in an acid fine sandy loam, pH 5.7. In a 6-wk study, MacLean and Dekker (1978) observed reduced yield at the lowest Ni concentration of 30 mg in a loam soil (Greville loam), whereas 480 mg Ni kg^{-1} was needed in an unlimed clay soil (Rideau clay). Liming the Rideau clay eliminated the effect of Ni. Gupta et al. (1987) also studied yield effects of Ni salts on lettuce and found no effects at 29 mg Ni kg^{-1} in a sandy loam (Steinhof) with low pH of 4.9. Effects were observed, however, after adding 46 mg Ni kg^{-1}. In less acidic soils (Gänsemos, Erlach, and Gasel), the highest no-effect levels were observed at 172–355 mg Ni kg^{-1}, with corresponding effect levels at 253–503 mg Ni kg^{-1}.

Corn.   Investigating the toxic effects of Ni on corn, Wallace et al. (1980) observed an 8% reduction in yield at 100 mg Ni kg^{-1} in a loam soil, although this was not significantly different from the control plants. Using a sandy soil, Metwally and Rabie (1989) observed reduced yield of corn exposed to 80 mg but observed no effct in an alluvial soil at 160 mg Ni kg^{-1}. Internal concentrations of Fe and P (increasing) and Zn and Mn (decreasing) were also affected by Ni. MacLean and Dekker (1978) investigated the effect of Ni on corn in two soils. In a Grenville loam, Ni affected the yield at 60 mg Ni kg^{-1}, whereas 240 mg was needed in an unlimed Rideau clay. Liming the Rideau clay reduced the toxic effect at 240 mg Ni kg^{-1}.

Soybeans, Mustard, Onion, Fenugreek, and Alfalfa.   A number of other plant species have also been studied, but even fewer reports are available on these species. Webber (1972) examined the effect of Ni on mustard and found a reduced yield (10%–30%) following addition of 50 mg Ni kg^{-1}. Roth et al. (1971) found that soybean yield was affected at 1879 mg Ni kg^{-1}, with no effects at 939 mg Ni kg^{-1}. Wallace et al. (1980) found 10% nonsignificant reduction upon addition of 100 mg Ni kg^{-1}. Studying *Fabia* beans, Metwally and Rabie (1989) observed no effect at 50 mg Ni kg^{-1} in alluvial or sandy soils. Dang et al. (1990) studied the effect of Ni and other metals on the growth of fenugreek and onion; adding 50 mg Ni kg^{-1} to a clay loam caused a significant 20% reduction in the yield of both onion and fenugreek, with a total failure at 400 mg Ni kg^{-1}. Halsted et al. (1969) studied the toxic effects of Ni in alfalfa and found no effects at 20 mg but did note effects at 50 mg Ni kg^{-1}. Using soils with higher organic content or higher pH, the lowest-effect level increased to 500 mg Ni kg^{-1}. Using two sandy loams, Taylor and Allinson (1981) found reduction in alfalfa yield after adding 125 mg, with more than 50% reduction at 250 mg Ni kg^{-1}. It should be noted that in this experiment the metal was not uniformly mixed but rather added to the top layer of soil to simulate aerial deposition.

Woody Species.   Patterson and Olson (1983) investigated radial elongation of seedlings for a range of woody species. In a mineral soil, white spruce (*Picea mariana*) was affected at addition of 50 mg Ni kg^{-1} (34% reduction), and for white pine (*Pinus strobus*) the no-observable-effect (NOEC) concentration was found at 50 mg and the lowest-observable-effect concentration (LOEC) was 100 mg Ni kg^{-1} (53% reduction). Similar but not significant reductions were found for jack pine (*Pinus resinosa*), paper birch (*Betula papyrifera*), and red pine (*Pinus resinosa*). No effects were observed at 1000 mg Ni kg^{-1} when the plants were grown in soils rich in organic matter.

D. Invertebrates

Few papers have reported the toxic effects of nickel on soil-dwelling inverte-brates, and the available studies all considered earthworm species (Table 6). Some studies of Ni as a contaminant of food were found; however, these are not reported here as it is impossible to equate these to soil concen-trations.

Studying both growth rate and mortality of *Lumbricus rubellus*, Ma (1982) observed no reduction in growth after 6 wk at 20 mg Ni kg^{-1} but a 50% decrease in growth rate after 12 wk. At 150 mg Ni kg^{-1}, growth rate was reduced by 80%. No mortality was observed until much higher concentrations were reached, with LC_{50} values between 2000 and 2500 mg Ni kg^{-1}. In a 12-wk study on *L. rubellus*, Ma (1982; as cited by van de Meent et al. 1990) observed the no-effect concentration for reproduction to be 50 mg and that for growth to be 85 mg Ni kg^{-1}.

Studying the toxic effects of Ni on the earthworm *Eisenia fetida* in the laboratory, Neuhauser et al. (1983) used a two-layer experiment with metals only in the top layer, which consisted of manure. In a short-term exposure (8 wk), the lowest concentration causing a growth reduction was 500 mg, with a corresponding no-observable effect at 300 mg Ni kg^{-1}. For reproduc-tion, effects were observed at 300 mg, with the no-observable effect concen-tration at 100 mg Ni kg^{-1}. In a longer term study (20 wk), growth was not inhibited at 400 mg Ni kg^{-1}, but reproduction was reduced at this level. Neuhauser et al. (1984) also studied the effect of Ni in a two-layer test similar that just described; both growth and reproduction were affected at 250 mg Ni kg^{-1} in a 6-wk exposure. In a 12-wk test, growth was reduced only at or above 500 mg, whereas reproduction was affected at 250 mg Ni kg^{-1}. Neuhauser et al. (1985, 1986) obtained an LC_{50} of 757 mg Ni kg^{-1}, with a standard deviation of 661–867 mg Ni kg^{-1}, in an artificial soil test (OECD soil). Several different Ni salts were tested, but no significant dif-ferences in their toxicity were observed.

Hartenstein et al. (1981) exposed *E. fetida* for 8 wk to heavy metals as sludge added as a top layer. Growth was inhibited at concentrations be-tween 1200 and 12,000 mg Ni kg^{-1}, at which concentrations mortality also occurred. In a similar experiment with a contaminated manure mixture placed on top of the soil, Malecki et al. (1982) investigated the effect of several different Ni salts. They found reduced growth following addition of 200 mg Ni kg^{-1} as $NiCl_2$, and at 500 mg Ni kg^{-1} for the acetate, carbonate, nitrate, and sulfate salts. Nickel oxide caused no effect at 40,000 mg Ni kg^{-1}. A similar picture was obtained for reproduction data, although the carbonate salt was four times less sensitive than found for growth. It should be noted that in these studies using a two-layered design, with Ni added only to the top layer, the worms may have avoided the contaminated layer and thus the toxicity of Ni may have been underestimated.

J.J. Scott-Fordsmand

Table 6. NOEC, LOEC, and EC_x values for effects of Ni on invertebrates in soil.

Reference	Soil	Compound	Time *	Parameter	Species	Effect data (mg Ni kg^{-1} dry wt soil)	
						NOEC	LOEC/EC$_{(x)}$
Hartenstein et al. (1981)	Silt loam/sludge	NiSO$_4$	8 wk	Growth	Earthworm (Eisenia fetida)		1200–12000
				Reproduction			1200–12000
Ma (1982)	Sandy loam	NiCl$_2$	6 wk	Mortality	Earthworm (Lumbricus rubellus)		2000–2500$_{(50)}$ E
							20$_{(50)}$ E
			6–12 wk	Growth			37 T
Ma (1982; as cited by van de Meent et al. 1990)	—	NiCl$_2$	12 wk	Reproduction	Earthworm (Lumbricus rubellus)	50	
				Growth		85	
Malecki et al. (1982)	—/sludge	Different Ni salts		Growth	Earthworm (Eisenia fetida)		200–500
				Reproduction			200–2000
Neuhauser et al. (1983, 1984)	—/sludge	Different Ni salts	6–12 wk	Growth	Earthworm (Eisenia fetida)	250	500
				Reproduction		250–500	500–1000
Neuhauser et al. (1985, 1986)	OECD	Ni(NO$_3$)$_2$	2 wk	Mortality	Earthworm (Eisenia fetida)		757$_{(50)}$ N

OECD: sand = 69%, clay = 10%, organic matter = 20%, CaCO$_3$ = 7%, pH = 6.9.

IX. Ecological Risk of Nickel in Denmark

As discussed earlier, there is evidence that Ni is toxic to soil-dwelling organisms. However, to evaluate the ecological risk of Ni in the terrestrial environment one also needs, apart from toxicity, to consider the bioavailability of Ni and its possible bioaccumulation and biomagnification.

Nickel is generally uniformly distributed through the soil profile but may accumulate in the top layer as a result of deposition or from decomposition of living organisms. In Denmark, soil concentrations outside urban areas, in general, range from 0.5 to 15 mg Ni kg^{-1}, with a mean of 6 mg Ni kg^{-1}. The soil Ni concentration is highly dependent on the soil texture, with a high Ni concentration in clay soils and a low concentration in sandy soils. Within the soil, Ni may exist in several different forms, depending on the soil type and other physicochemical conditions. Nickel compounds, of which Ni^{2+} is the most dominant, are in general rather soluble at a pH below 6.5 but insoluble and slowly mobile at higher pHs. Because Ni has increased solubility below pH 6.5, it may be mobilized by acidification (acid rain).

The bioavailability and the bioaccumulation of Ni also depend on the prevailing physicochemical conditions, but most organisms do not accumulate Ni to concentrations above the ambient soil concentrations, although some organisms do so. Special hyper-accumulating organisms may contain considerable amounts of Ni. At present, it is not known whether Ni is biomagnified, but the low bioconcentration factors and rather rapid excretion by earthworms indicate that this is not the case.

Several studies have shown that Ni is toxic to microorganisms and microbial processes. In a few cases, the toxicity of Ni to microorganisms begins at soil concentrations of 10 mg Ni kg^{-1}, indicating an even lower no-effect level, but in most cases the effects are observed at concentrations of 30–50 mg Ni kg^{-1} or more. For plants the toxicity effects of nickel, in a few cases, also begin at soil concentrations about 10–12.5 mg, with corresponding no-observable-effect levels of 6 mg, but in most cases are greater than 20–30 mg Ni kg^{-1}. For invertebrates, experiments showed effects starting at 37 mg Ni kg^{-1}, but in one case an EC_{50} value of 20 mg Ni kg^{-1} was extrapolated for earthworms. For invertebrates, it should be kept in mind that only earthworm species were examined, giving no indication of the toxicity toward such arthropods as mites and Collembola or other groups of soil-dwelling macro- or microorganisms.

These results suggest that effects may occur at soil concentrations starting at 10 mg Ni kg^{-1}, for the sensitive species, but in most cases effects do not begin until 25–50 mg Ni kg^{-1}. On this basis, and depending on background concentrations, the Danish soil quality criterion was set at 10 mg Ni kg^{-1}. Other countries have elaborated other guidelines or quality criteria for Ni in soil, based on different concepts and policies and on the different

Table 7. Current risk assessment criteria for nickel in soils in selected countries.

Country	Ni (mg kg^{-1} dry wt)
Denmark (quality criterion) (Scott-Fordsmand and Pedersen 1995)	10
Germany (multifunctionality II) (Visser 1993)	50
The Netherlands (target value) (VROM 1994)	35
Switzerland (most sensitive purposes) (Visser 1993)	30

national background levels of Ni in their soil (Table 7). It is important to realize that the results presented herein are based on experiments with addition of soluble salts, which makes Ni more available than would be expected in most field situations. How much more available Ni is in laboratory experiments compared to field situations is presently impossible to estimate. A comparison of the toxicity levels obtained from the literature with the Ni levels found outside urban areas in Denmark, a mean of 6 mg Ni with a range of 0.5 to 15 mg Ni kg^{-1}, indicates that Ni does not at present seem to be of major concern. This is further supported by the good correlation between the soil texture (especially the clay fraction) and soil Ni concentrations, which indicates that a large part of the Ni found originates from natural sources, as shown in Fig. 1 and Section IV.

Although Ni does not seem to be of major concern outside urban areas, precautions should be implemented as there is a lack of information on its long-term effects on the soil ecosystem. Furthermore, Ni may in certain areas act as a chemical time bomb because of reduction in pH. For example, altered agricultural practices, such as reduction of liming, may delay the occurrence of harmful effects caused by Ni mobilization under reduced pH. It should also be noted that the impact of Ni when present in the environment in a mixture with other pollutants is only poorly understood. Finally, Ni may present a major problem in land near towns, in industrial areas, or even in agricultural land receiving wastes such as sewage sludge and pig manure (Larsen et al. 1996).

Summary

Nickel is toxic to all soil-dwelling organisms at some level. Nickel is generally uniformly distributed through the soil profile but typically accumulates at the surface from deposition by man's industrial and agricultural activities. In Denmark, mean soil concentrations outside urbanized areas are 6 mg Ni kg^{-1}, ranging from 0.5 to 15 mg, higher in clay and lower in sandy soils. Nickel compounds are fairly soluble at pH 6.5 and lower, so acidification may be a significant factor in its soil mobilization.

The majority of research reports used in this review are based on experiments involving the addition of soluble nickel salts to soils, making Ni more available than would be expected in field situations. Thus, it is difficult to estimate how much more Ni is available in these laboratory experiments than in the field. Most organisms do not accumulate Ni to levels above that of the ambient soil concentration, though there are several exceptions. It is toxic to plants, microorganisms, and microbial processes, beginning at levels of 10 mg Ni kg^{-1}, but in most instances effects are not observed until levels of 30–50 mg Ni kg^{-1} are reached. Earthworms, apparently the only invertebrates studied, begin to show reproductive and growth effects at about 37 mg kg^{-1}.

Based on these and related studies, the Danish soil quality criterion was set at 10 mg kg^{-1} Ni. Other countries have established other guidelines and criteria based on different concepts and policies and on different national background levels. The impact of Ni when present in the environment in a mixture with other pollutants is neither well researched nor well understood. Nickel apparently does not seem to be of major concern outside urban areas at this time but may eventually become a problem as a result of decreased soil pH caused by reduced use of soil liming in agriculture and mobilization as a consequence of increased acid rain.

Acknowledgments

The author thanks Dr. J. Bak for providing the map of Denmark, and Dr. H. Løkke and Dr. S.P. Hopkin for their critical comments on drafts of this review.

References

Adriano DC (1986) Trace Elements in the Terrestrial Environment. Springer-Verlag, New York.

Al-Khafaji AA, Tabatabai MA (1979) Effects of trace elements on arylsulfatase activity in soils. Soil Sci 127(3):129–133.

Allison DW, Dzialo C (1981) The influence of lead, cadmium, and nickel on the growth of ryegrass and oats. Plant Soil 62:81–89.

Århus Amt (1992) Tungmetaller i Århus Amt. Baggrundsværdier i tertiære sedimenter og recente jordbunde. Miljøkontoret, Århus Amt.

Babich H, Stotzky G (1982) Toxicity of nickel to microorganisms in soil: influence of some physicochemical characteristics. Environ Pollut 29(A):303–315.

Beck T (1981) Untersuchungen über die toxische wirkung der in siedlungsabfällen häufigen schwermetalle auf die bidenmikroflora. Z Pflanzenernaehr Dueng Bodenkd 144:613–617.

Beyer NW, Chaney RL, Mulhern BM (1982) Heavy metal concentrations in earthworms from soil amended with sewage sludge. J Environ Qual 11(3):381–385.

Bhuiya MRH, Cornfield AH (1972) Effects of addition of 1000 ppm Cu, Ni, Pb, and Zn on carbon dioxide release during incubation of soil alone and after treatment with straw. Environ Pollut 3:173–177.

Bingham FT, Page AL, Mitchell GA, Strong JE (1979) Effects of liming an acid soil amended with sewage sludge enriched with Cd, Cu, Ni, and Zn on yield and Cd content of wheat grain. J Environ Qual 8:202–207.

Boyd RS, Martens SN (1994) Nickel hyperaccumulated by *Thlaspi montanum* var. *montanum* is acutely toxic to an insect herbivore. Oikos 70:21–25.

Bremner JM, Douglas LA (1971) Inhibition of urease activity in soils. Soil Biol Biochem 3:297–307.

Brooks NN (1980) Accumulation of nickel by terrestrial plants. In: Nriagu JO (ed) Nickel in the environment. Wiley & Sons, New York, pp 407–430.

Cataldo DA, Garland TR, Wildung RE (1978) Nickel in plants I. Uptake kinetics using intact soybean seedlings. Plant Physiol 62:563–565.

Chander K, Brookes PC (1991) Effect of heavy metals from past applications of sewage sludge on microbial biomass and organic matter accumulation in sandy loam and silty loam U.K. soils. Soil Biol Biochem 23:927–932.

Chaudri AM, McGrath SP, Giller KE (1992) Survival of indigenous population of *Rhizobium leguminosarum* biovar *trifolii* in soil spiked with Cd, Zn, Cu and Ni salts. Soil Biol Biochem 24:625–632.

Chaudri AM, McGrath SP, Giller KE, Rietz E, Sauerbeck DR (1993) Enumeration of indigenous *Rhizubium leguminosarum* biovar *trifolii* in soils previously treated with metal-contaminated sewage sludge. Soil Biol Biochem 25:301–309.

Cornfield AH (1977) Effects of addition of 12 metals on carbon dioxide release during incubation of an acid soil. Geoderma 19:199–203.

Dahiya DJ, Singh JP, Kumar V (1994) Nitrogen uptake in wheat as influenced by the presence of nickel. Arid Soil Res Rehabil 8:51–58.

Dang YP, Chabra R, Verma KS (1990) Effect of Cd, Ni, Pb, and Zn on growth and chemical composition of onion and fennugreek. Commun Soil Sci Plant Anal 21(9–10):717–735.

De Catanzaro JB, Hutchinson TC (1985) Effect of nickel addition on nitrogen mineralization, nitrification and nitrogen leaching in some boreal forest soils. Water Air Soil Pollut 24:153–164.

de Haan S, Rethfeld H, van Driel W (1985) Acceptable levels of heavy metals (Cd, Cr, Cu, Ni, Pb, Zn) in soils, depending on their clay and humus content and cation-exchange capacity. Rapport 9-85, Institute voor Bodemvruchtbaarheid, Haren, The Netherlands.

Den Danske Jordklassificering (1976) Teknisk redegørelse. Landbrugsministeriet, Copenhagen, Denmark.

Doelman P, Haanstra L (1984) Short-term and long-term effects of cadmium, chromium, copper, nickel, lead, and zinc on soil microbial respiration in relation to abiotic soil factors. Plant Soil 79:317–327.

Doelman P, Haanstra L (1986) Short- and long-term effects of heavy metals on urease activity in soils. Biol Fertil Soils 2:213–218.

El-Sharouny HMM, Bagy MM, El-Shanawany AA (1988) Toxicity of heavy metals to Egyptian soil fungi. Int Biodeterior 24:49–64.

Frankenberger WT, Tabatabai MA (1981) Amidase activity in soils: IV. Effects of trace elements and pesticides. Soil Sci Soc Am J 45:1120–1124.

Frankenberger WT, Tabatabai MA (1991) Factors affecting L-asparaginase activity in soils. Biol Fertil Soils 11:1–5.

Friborg R (1992) Tungmetalprojektet i Sønderjyllands Amt. Grundvandsafde-
    lingen, Sønderjyllands Amt, Denmark.
Frossard R, Stadelmann FX, Niederhauser J (1989) Effects of different heavy met-
    als on fructan, sugar and strach content of ryegrass. J Plant Physiol 134:180–185.
Frostegård Å, Tunlid A, Bååth E (1993) Phospholipid fatty acid composition, bio-
    mass, and activity of microbial communities from two soil types experimentally
    exposed to different heavy metals. Appl Environ Microbiol 59(11):3605–3627.
Fu MH, Tabatabai MA (1989) Nitrate reductase activity in soils: effects of trace
    elements. Soil Biol Biochem 21:943–946.
Giashuddin M, Cornfield AH (1978) Incubation study on effects of adding varying
    levels of nickel (as sulphate) on nitrogen and carbon mineralisation in soil.
    Environ Pollut 15:231–234.
Giashuddin M, Cornfield AH (1979) Effect of adding nickel (as oxide) to soil on
    nitrogen and carbon mineralisation at different pH values. Environ Pollut 19:
    67–70.
Gish CD, Christensen RE (1973) Cadmium, nickel, lead, and zinc in earthworms
    from roadside soil. Environ Sci Technol 7(11):1060–1062.
Grün M, Machlette B, Kronemann H, Martin M, Schneider J, Podlesak W (1994)
    Übersichten Tierernahrung 22(1):7–16.
Gupta SK, Hani H, Santschi E, Stadelmann FX (1987) The effect of graded doses
    of nickel on the yield, the nickel content of lettuce and the soil respiration.
    Toxicol Environ Chem 14:1–10.
Haanstra L, Doelman P (1984) Glutamic acid decomposition as a sensitive measure
    of heavy metal pollution in soil. Soil Biol Biochem 16:595–600.
Haanstra L, Doelman P (1991) An ecological dose-response model approach to
    short- and long-term effects of heavy metals on arylsulphatase activity in soil.
    Biol Fertil Soils 11:18–23.
Halstead RL, Finn BJ, MacLean AJ (1969) Extractability of nickel added to soils
    and its concentrations in plants. Can J Soil Sci 49:335–342.
Haq AU, Bates TE, Soon YK (1980) Comparison of extractants for plant-available
    zinc, cadmium, nickel, and copper in contaminated soils. Soil Sci Soc Am J 44:
    772–777.
Hartenstein R, Neuhauser EF, Collier J (1980) Accumulation of heavy metals in the
    earthworm *Eisenia foetida*. J Environ Qual 9:23–26.
Hartenstein R, Neuhauser EF, Narahara A (1981) Effects of heavy metal and other
    elemental additives to activated sludge on growth of *Eisenia foetida*. J Environ
    Qual 10(3):372–376.
Hattori H (1992) Influence of heavy metals on soil microbial activities. Soil Sci
    Plant Nutr 38:93–100.
Hausinger RP (1992) Nickel and human health: current perspectives. In: Nieboer E,
    Nriagu, JO (eds) Biological Utilization of Nickel. Wiley & Sons, New York, pp
    21–36.
Hopkin SP (1989) Ecophysiology of Metals in Terrestrial Invertebrates. Elsevier
    Applied Science, New York.
IPCS (1992) Environmental health criteria 108: nickel. World Health Organization
    (WHO), Geneva.
Janssen MPM, Ma WC, van Straalen NM (1993) Biomagnification of metals in
    terrestrial ecosystems. Sci Total Environ (Suppl):511–524.
Juma NG, Tabatabai MA (1977) Effects of trace elements on phosphatase activity
    in soils. Soil Sci Soc Am J 41:343–346.

Kabata-Pendias A, Pendias H (1984) Trace Elements in Soils and Plants. CRC Press, Boca Raton, FL.

Khalid BY, Tinsley J (1980) Some effects of nickel toxicity on rye grass. Plant Soil 55:139–144.

Korte NE, Skopp J, Fuller WH, Niebla EE, Alesii BA (1976) Trace element movement in soils: influence of physical and chemical properties. Soil Sci 122: 350–359.

Krämer U, Cotter-Howells JD, Charnock JM, Baker AJM, Smith JAC (1996) Free histidine as a metal chelator in plants that accumulate nickel. Nature 379: 635–638.

Larsen MM, Bak J, Scott-Fordsmand JJ (1996) Monitering af tungmetaller i dansk dyrknings-og naturjord. Faglig Rapport Fra DMU Nr. 157, Danmarks Miljøundersgelser, Denmark.

Lepp NW (1980) Effect of Heavy Metal Pollution on Plants. Elsevier Applied Science, New York.

Liang CN, Tabatabai MA (1977) Effects of trace elements on nitrogen mineralization in soils. Environ Pollut 12:141–147.

Liang CN, Tabatabai MA (1978) Effects of trace elements on nitrification in soils. J Environ Qual 7:291–293.

Lighthart B, Baham J, Volk VV (1983) Microbial respiration and chemical speciation in metal amended soils. J Environ Qual 12:543–548.

Ma WC (1982) Biomonitoring of soil pollution: ecotoxicological studies of the effect of soil-borne heavy metals on lumbricoid earthworms. Annual report, Research Institute for Nature Management, Arnhem, The Netherlands, pp 83–97.

MaClean AJ, Dekker AJ (1978) Availability of zinc, copper and nickel to plants grown in sewage-treated soils. Can J Soil Sci 58:381–389.

Malecki MR, Neuhauser EF, Loehr RC (1982) The effect of metals on the growth and reproduction of *Eisenia foetida* (Oligochaeta, Lumbricidae). Pedobiologia 24:129–137.

McIlveen WD, Negusanti JJ (1994) Nickel in the terrestrial environment. Sci Total Environ 148:109–138.

Merian E (1984) Introduction on environmental chemistry and global cycles of chromium, nickel, cobalt, beryllium, arsenic, cadmium, and selenium, and their derivates. Toxicol Environ Chem 8:9–38.

Metwally AI, Rabie MH (1989) Effect of nickel addition on plant growth and nutrient uptake in two soils. Egypt J Soil Sci 3:261–274.

Miljøstyrelsen (1985) Forbrug og forurening med arsen, chrom, cobalt og nikkel. Orientering fra Miljøstyrelsen, nr. 7, Copenhagen, Denmark.

Miljøstyrelsen (1995) Notat vedrøorende spildevandslam oktober 1994. Miljøstyrelsen, Hav-og Spildevandskontoret, Copenhagen, Denmark.

Mishra D, Kar M (1974) Nickel in plant growth and metabolism. Bot Rev 40:395–452.

Mitchell GA, Bingham FT, Page AL (1978) Yield and metal composition of lettuce and wheat on soils amended with sewage sludge enriched with cadmium, copper, nickel, and zinc. J Environ Qual 7:165–170.

Neuhauser EF, Malecki MR, Loehr RC (1983) Methods using earthworms for the evaluation of potentially toxic materials in soils. Hazardous and industrial solid waste testing. In: Conway RA, Gulledge WP (eds) Second symposium ASTM STP, American Society for Testing and Materials, Philadelphia, PA, pp 313–320.

Neuhauser EF, Malecki MR, Loehr RC (1984) Growth and reproduction of the earthworm *Eisenia fetida* after exposure to sublethal concentrations of metals. Pedobiologia 27:89–97.

Neuhauser EF, Loehr RC, Milligan DL, Malecki MR (1985) Toxicity of metals to the earthworm *Eisenia fetida*. Biol Fert Soil 1:149–152.

Neuhauser EF, Loehr RC, Malecki MR (1986) Contact and artificial soil tests using earthworms to evaluate the impact of waste in soil. In: Petros JK, Lacy WJ, Conway RA (eds) Hazardous and industrial solid waste testing. Fourth symposium, ASTM STP. 886, American Society for Testing and Materials, Philadelphia, PA.

Neuhauser EF, Cukic ZV, Malechi MR, Loehr RC, Durkin PR (1995) Bioconcentration and biokinetics of heavy metals in the earthworm. Environ Pollut 89:293–301.

Nieboer E, Richardson DHS (1980) The replacement of the nondescriptive term 'heavy metals' by a biologically and chemically signifcant classification of metal ions. Environ Pollut 1:3–26.

Patterson WA, Olson JL (1983) Effects of heavy metals on radical growth of selected woody species germinated on filter paper, mineral, and organic soil substrates. Can J For Res 13:233–238.

Rogers JE, Li SW (1985) Effects of metals and other inorganic ions on soil microbial activity: soil dehydrogenase assay as a simple toxicity test. Bull Environ Contam Toxicol 34:858–865.

Roth JA, Wallihan EF, Sharpless RG (1971) Uptake by oats and soybeans of copper and nickel added to a peat soil. Soil Sci 112:338–342.

Sadiq M (1985) Uptake of cadmium, lead and nickel by corn grown in contaminated soils. Water Air Soil Pollut 26:185–190.

Schmitt HW, Sticher H (1992) Heavy metal compounds in soil. In: Merian E (ed) Metals and Their Compounds in the Environment. VCH Publishers, New York, pp 311–331.

Scott-Fordsmand JJ, Pedersen MB (1995) Soil quality criteria for selected inorganic compounds. Working report (Arbejdsrapport) no. 48, Danish Environmental Protection Agency, Copenhagen, Denmark.

Singh BR, Jeng AS (1993) Uptake of zinc, cadmium, mercury, lead, chronium, and nickel by ryegrass grown in a sandy soil. Norw J Agric Sci 7:147–157.

Sorteberg S (1974) The effect of some heavy metals on oats in a pot experiment with three different soil types. J Sci Agric Soc Finl 3:277–288.

Spalding BP (1979) Effects of divalent metal chlorides on respiration and extractable enzymatic activities of Douglas-fir needle litter. J Environ Qual 8:105–109.

Stadelmann FX, Santschi-Fuhrimann E (1987) Beitrag zur Abstuetzung von Schwermetallrichtwerten im Boden mit Hilfe von Bodenatmessungen. FAC Liebefeld, Switzerland.

Stott DE, Dick WA, Tabatabai MA (1985) Inhibition of pyrophosphatase activity in soils by trace elements. Soil Sci 139:112–117.

Sunderman FW, Oskarsson E (1991) Nickel. In: Merian E (eds) Metals and Their Compounds in the Environment. Occurence, Analysis, and Biological Relevance. VCH Publishers, New York pp 1101–1126.

Tabatabai MA (1977) Effects of trace metals on urease activity in soils. Soil Biol Biochem 9:9–13.

Taylor RW, Allinson DW (1981) Influence of lead, cadmium, and nickel on the growth of *Medicago sativa* (L.). Plant Soil 60:223–236.

Tjell JC, Hovmand MF (1978) Metal concentrations in Danish arable soils. Acta Agric Scand 28:81–89.

Tyler G, Schlüter K, Rühling Å, Rundgren S (1992) Mercury in soil — distribution, speciation, and biological effects. NORD 1992:3.

Tyler LD, McBride MB (1982) Influence of Ca, pH, and humic acid on Cd uptake. Plant Soil 64:259–262.

Uren NC (1992) Forms, reactions, and availability of nickel in soils. Adv Agron 48: 141–203.

van de Meent D, Aldenberg T, Canton JH, van Gestel CAM, Sloff W (1990) Desire for levels. Background study for the policy document "Setting Environmental Quality Standards for Water and Soil." RIVM annex to rep. 670101 002. National Institute of Public Health and Environmental Protection, Bilthoven, The Netherlands.

Visser WJF (1993) Contaminated land policies in some industrialized countries. Technische Commissie Bodemdescherming R02, The Hague, The Netherlands.

VROM (1994) Environmental quality objectives in the Netherlands. A review of environmental quality objectives and their policy framework in the Netherlands. Risk Assessment and Environmental Quality Division, Directorate for Chemicals, External Safety and Radiation Protection, Ministry of Housing, Spatial Planning and the Environment, The Hague, The Netherlands.

Wallace A, Romney EM, Mueller RT, Lunt OR (1980) Influence of environmental stresses on response of bush bean plants to excess copper. J Plant Nutr 2:39–49.

Webber J (1972) Effects of toxic metals in sewage on crops. Water Pollut Control 71:404–413.

Welch RM (1995) Micronutrient nutrition of plants. CRC Crit Rev Plant Sci 14(1): 49–82.

Whitby LM, Gaynor J, Maclean AJ (1978) Metals in soils of some watersheds in ontario. Can J Soil Sci 58:325–330.

Wild H (1975) Termites and the serpentines of the great dyke of Rhodesia. Trans Rhod Sci Assoc 57(1):1–11.

Wilke BM (1988) Langzeitwirkungen potentieller anorganischer Schadstoffe auf die mikrobielle Aktivität einer sandigen Braunerde. Z Pflanzenernähr Dueng Bodenkd 151:131–136.

Yadav DS, Kumar V, Singh M (1986) Inhibition of soil urease activity and nitrification with some metallic cations. Aust J Soil Res 24:527–532.

Manuscript received April 23, 1996; accepted April 25, 1996.

Effects of Acidic Deposition on Soil Invertebrates and Microorganisms

Roman G. Kuperman* and Clive A. Edwards**,†

Contents

I. Introduction
A. History of Acidic Deposition

The first observations of the detrimental effects of atmospheric pollution, including acidic deposition, on plants and other organisms were made in Europe more than 300 yr ago. However, scientific research into these effects did not begin until 1852, when Robert Smith, an English chemist, described the rain chemistry in the vicinity of Manchester, England. This study and other work that contributed to our understanding of the historical development of scientific knowledge and public concern over acidifying compounds in precipitation have been reviewed by Cowling (1982).

---

*U.S. Army Edgewood Research, Development & Engineering Center, SCBRD RTL Bldg. E-3220, Aberdeen Proving Ground, MD 21010-5423, U.S.A.

**Department of Entomology, Ohio State University, Columbus, OH 43210, U.S.A.

†Corresponding author.

© 1997 by Springer-Verlag New York, Inc.

Reviews of Environmental Contamination and Toxicology, Vol. 148.

Eville Gorham (1955) suggested that soils, streams, and lakes could be acidified by chemicals in the rain that derived from industrial emissions. In the 1950s, the European Air Chemistry Network (EACN) was initiated by the University of Stockholm and the Swedish University of Agricultural Sciences, Uppsala, to characterize the long-term changes in precipitation chemistry associated with human activities. This network was expanded later to other European countries. By the early 1970s, it became apparent to Swedish researchers that acidic deposition was leading to acidification of some ecosystems and, moreover, that it was a global phenomenon. In 1980, the Swedish network was restructured to form part of a national monitoring program organized by the National Swedish Environment Protection Board. Some of these sites participated in international programs organized by the United Nations Environmental Program (UNEP). In addition to the data provided by the UNEP, a large amount of information on precipitation chemistry is now available from the European Monitoring and Evaluation program (EMEP).

A historical overview of European wet deposition monitoring and assessment of the amounts of SO_4^{2-} in precipitation for the period 1955–1982 was provided by Rodhe and Granat (1984). Their important conclusions suggested that in some countries, including Norway, Sweden, Denmark, and Finland, the median excess SO_4^{2-} concentrations increased by approximately 50% from the late 1950s to the late 1960s. A further increase during the 1970s was indicated by the Danish stations, but most of the stations in Sweden and Norway reported an average decline of about 20% by the late 1970s. The SO_4^{2-} data from the United Kingdom and continental Europe varied considerably with few overall trends. The average levels of sulfates and acidity in precipitation were generally higher in Europe than in North America in the 1970s, but the European values declined sharply by the mid-1980s. This reduction in amounts of sulfates in precipitation was attributed to a reduction of overall emission sources in Europe (NAPAP 1990).

Wet deposition monitoring in Canada began in the early 1970s, mainly as individual efforts rather than as a coordinated national program (NAPAP 1990). Prior to 1970, data on acidic precipitation were collected to support scattered studies of nutrient budgets or smelter damage in localized geographic areas. Concern about nutrient inputs led to the establishment of a monitoring network around the Great Lakes in 1961. As concern grew about the effects of acid rain resulting from point-source emissions, more monitoring sites were established, particularly in Ontario and Alberta. A historical overview of precipitation and air quality monitoring in Canada was made by Whelpdale and Barrie (1982). Data collected for the period 1976–1985 (Dillon et al. 1988) showed that concentrations of sulfate and hydrogen ions in precipitation had decreased by about 30% from the first half of this study period to the last half, while nitrate concentrations decreased by only about 10%. These results are consistent, for the most part, with those reported from the northeastern United States for the same time period (NAPAP 1990).

In the U.S., concern about the effects of acidic deposition developed in the 1970s based on the findings by some American scientists of the highly acidic nature of precipitation in the northeastern U.S. (Likens and Borman 1974). In 1979, a 10-yr monitoring program to determine the causes and impact of acidic deposition in the U.S. was initiated (NAPAP 1990). Although numerous reports of wet chemistry-based data on acid precipitation are available from one or more sites for periods ranging from 1 to 5 yr (Zemba et al. 1988), few are related to longer time intervals. An exception was that from Hubbard Brook (Galloway et al. 1984), where continuous deposition monitoring data records for several decades were established. A few reviews have summarized results from a monitoring network back into the 1970s (Munger and Eisenrich 1983; Wisniewski and Keitz 1983), but on a scale too coarse for inferring local site-specific patterns. The paucity of long-term acidic deposition records, particularly where diminished air quality has existed for the past half-century or more, creates a major problem for analysis of its potential effects when the likely response time extends back beyond network monitoring records (Armentano and Loucks 1990). Significant reductions in sulfur and nitrogen emissions from industrial sources in the U.S., particularly in the lower midwest region, have occurred over the past two decades, leading to a decreasing trend in the concentrations of acidic substances in precipitation (NAPAP 1987). This decline has been attributed to the increasing use of low-sulfur coal by power plants, installation of pollution control devices such as acidic gas scrubbers, and the closing of older power plants.

B. Types and Sources of Acidic Deposition

The term acidic deposition refers to the total amount of acids and acidifying compounds present in all forms of both wet (rain, snow, fog, mist) and dry (particulate, aerosols, gases) deposition. Deposition is usually considered to be acidic if its pH is less than 5.6. This pH value occurs for water in equilibrium with the carbon dioxide normally present in the atmosphere. Wet acidic deposition is composed essentially of sulfuric and nitric acids, whereas dry deposition of acidic sulfur compounds occurs usually as SO_2, which is oxidized further to sulfuric acid after deposition. Analyses of U.S. regional deposition data for sulfur and NO_x nitrogen indicate that wet deposition and dry deposition of anthropogenic sulfur over the U.S. and Canada are of almost equal importance, containing 30% of sulfur and 29% of nitrogen emissions, respectively. Droplet deposition has been equivalent to only about 4% of emissions, and 37% of the total anthropogenic sulfur emissions are exported from the region (Shannon and Sisterson 1992). For NO_x nitrogen, corresponding emission estimates are 43% dry and 30% wet, with 4% of emissions as droplets and 23% as net export (Shannon and Sisterson 1992).

Most sulfur oxides are produced by combustion of sulfur-containing fuels or from sulfur contained in metal ores that are being extracted. Some types of fuels contain nitrogen that can be oxidized by oxygen in the air when the fuel is burned, leading to formation of nitrogen oxides. Nitrogen oxides can also be formed from the nitrogen and oxygen from the air in many high-temperature fossil fuel combustion systems, even when the fuel contains little or no nitrogen. The oxidation of nitrogen results in the formation of two different nitrogen oxides: nitric oxide (NO) and nitrogen dioxide (NO_2). Both are reactive in the atmosphere and so can contribute to acidic deposition.

The main components of acidic deposition, such as sulfur dioxide, nitrogen oxides, volatile organic compounds (VOCs), and the primary compounds involved in acid–base chemistry, (e.g., ammonia, alkaline dust particles, primary sulfates, hydrogen chloride, and hydrogen fluoride), can originate from both natural and anthropogenic sources. According to the NAPAP (1990) report, more than 9000 industrial plants in the U.S. and Canada in 1985 had annual emissions in excess of 100 tons of SO_2 or NO_x. There are also emissions from more than 130 million U.S. passenger cars and numerous other small point sources that are very difficult to identify individually. Natural sources of acidifying compounds in atmospheric deposition include emissions from the oceans, volcanoes, and geysers and the vapor from trees, other plants, and decaying organic matter.

Dignon (1992) estimated that global emissions of nitrogen oxides from fossil fuel combustion only were about 22 Mt (N) yr^{-1} for 1980. The concurrent sulfur oxides emissions were about 62 Mt (N) yr^{-1}. Most of the emissions are produced in the Northern Hemisphere, including 95% of the nitrogen oxide and 94% of the sulfur oxide. The five countries with greatest estimated annual emissions of SO_x in Tg of S (Tg = 10^6 metric tons) in 1980 were the former Soviet Union (12.8), followed by the U.S. (12.4), China (7.8), Poland (2.7), and West Germany (2.2 Tg S) (Dignon 1992). The five countries with the largest estimated emissions of NO_x were the U.S. (6.4), followed by the U.S.S.R. (4.4), China (1.7), Japan (0.8), and West Germany (0.66 Tg N) (Dignon 1992).

Estimates of total annual acidic emissions from the U.S. in 1985 were 23.1 million tons of SO_2, 20.5 million tons of NO_x, and 22.1 million tons of VOCs (NAPAP 1990). This inventory indicated that over 75% of U.S. SO_2 emissions originate from large industrial plants, mostly electric utility plants that emit 10,000 tons or more annually; only 27% of national NO_x emissions came from such large industrial plants. Thus, the majority of SO_2 emissions come from a relatively small number of sites, usually in rural areas, where the power plants are located. A large proportion of NO_x emissions come from smaller, more dispersed point sources, such as motor vehicles, which are concentrated mainly in urban areas. The NAPAP report (1990) estimated that approximately 11% of the total (natural plus anthropogenic) emissions of nitrogen oxides come from natural biogenic sources, as well as lightning, while contribution of natural sources to sulfur compound emission was estimated to be between 1% and 5% of the total U.S. sulfur emissions. The report concluded that it is unlikely that naturally produced oxides of nitrogen contribute significantly to nitrate deposition over most of North America. Similarly, low levels of sulfur emissions from natural biogenic sources do not appear to make a significant contribution to the sulfate component of acidic deposition in the region.

C. Types and Mechanisms of Acidic Deposition Impact on Soil Ecosystems

Soil ecosystems have long been exposed to a variety of environmental perturbations, including those resulting from drought, fire, flood, and human activities. Most soil ecosystems are capable of adapting to such stresses in the long term. However, although most soil ecosystems are not very fragile, the extent of tolerance of some soils to some environmental stress factors can be relatively limited. To forecast accurately the consequences of various perturbations on soil ecosystems, it is necessary to understand the mechanisms of their effects.

The varied effects of acidic deposition on soil ecosystems are caused by the interactions of its diverse components with the very complex functional structure of the soil ecosystem. Physicochemical changes in soils exposed to acidic deposition, as well as natural soil processes that lead to their acidification, have been considered at length in numerous reports (Andersson et al. 1980; Binkley and Richter 1987; Burgess 1984; Driscoll and Likens 1982; Johnson and Siccama 1983; Loucks 1992; Nilsson et al. 1982; Postel 1984; Ulrich and Pankrath 1983; Voigh 1980). Therefore, only a brief review of the physicochemical effects of acidic deposition on soils is given here. The primary aim of this review is to describe and assess the extent of the effects (both direct and indirect) of these changes in acidity on the soil biota and on biologically mediated processes in soil.

The overall effect of acidic deposition on soil ecosystems can be either beneficial or detrimental depending on which physical and chemical properties of the soil are affected and the degree to which they are impacted. Addition of acid rain to alkaline soils can have a beneficial effect by lowering the soil pH to levels more optimal for plant growth. Increased deposition of nitrates and sulfates may also be favorable for plants because these are important plant nutrients, and increased acidity can also enhance the solubility of nutrients. However, in naturally acidic soils, additional deposition of acidic material may decrease the pH of soils to levels below the optimum range for growth and reproduction of plants and functioning of the soil decomposer community. Increases in total acidity of soil are usually associated with loss of base cations, reductions in cation-exchange capacity (CEC), increased solubility of potentially toxic metals such as aluminum and manganese, and reduction in the solubility of phosphorus and molybdenum, as well as mineral degradation.

Most hydrogen ions that are deposited or produced in soil are neutralized by reacting with one or more of the soil constituents. Ulrich and Pankrath (1983) distinguished several important processes that occur in soil with different pH ranges. Within a pH range of 6.2–8.6, H^+ ions are consumed by the dissolution of $CaCO_3$ (carbonic acid–carbonate neutralization range). In the pH range 5.0–6.2, H^+ ions are consumed by the weathering of silicates (carbonic acid–silicate neutralization range). The neutralization capacity in this case depends on the kind and amount of primary silicates present in soil. If strong acid (HNO_3) is produced at a rate exceeding the buffer rate, the pH may drop temporarily below 3.0. Within the pH range 4.2–5.0, H^+ ions are consumed by dissolving $AlOOH$ from weathered silicates with the formation of polymeric aluminum hydroxocations. These replace Ca^{2+} and Mg^{2+} at the surface of the cation exchangers (cation-exchange neutralization range). Within the pH range 3.0–4.2, H^+ ions are consumed by the formation of Al^{3+} ions from polymeric aluminum compounds (aluminum neutralization range). Moreover, within the pH range 3.0–3.5, H^+ ions are consumed by the formation of Fe ions from iron oxides (iron neutralization range).

As long as the H^+ ions that are produced or deposited can be neutralized, a soil ecosystem can maintain an equilibrium steady state. However, if the rate of deposition and production of H^+ ions exceeds the potential neutralization rate, the ecosystem will shift to the next buffer range. During this shift, it changes its composition, structure, and productivity (Overrein et al. 1981). If the rates of production and neutralization of H^+ ions in soil are equal, even a small addition of H^+ ions by deposition may induce such a shift, irrespective of the buffer range in which the ecosystem can exist (Last et al. 1980).

Cation-exchange capacity and base saturation are the most important factors contributing to the buffering capacity of soil. Cation exchange results from the negative electric charges on the clay and humus particles of the soil matrix. During this process, positively charged counterions such as Ca^{2+}, Mg^{2+}, K^+, Na^+, Mn^{2+}, and Al^{3+} can be adsorbed onto these particles, thus preserving electrical neutrality. The ability of H^+ and other cations to occupy the exchange sites follows LeChatelier's principle, i.e., if the concentration of a single ion is increased in the soil solution, that ion will tend to associate with the exchange complex. A decrease in the concentration of an ion in solution will result in displacement of that ion from the exchange site to the solution. If the concentrations of all exchangeable cations are equal, then the replacement series is Na > K > Mg > Ca > Al(H) (Levine and Ciolkosz 1988).

Base saturation is the fraction of the soil exchange complex that is occupied by cations such as Ca^{2+}, Mg^{2+}, K^+, and Na^+. An exchange complex dominated by the basic cations will have a soil solution with a relatively high pH because the exchange complex behaves as a dissociated acid. An exchange complex dominated by the acidic cations (H^+ and Al^{3+}) behaves as undissociated acids and will maintain a soil solution at lower pH values (Binkley et al. 1989). A major concern about inputs of acidic deposition to soil ecosystems is that soil exchange complexes may become progressively dominated by acidic cations as basic cations are leached from the soil, thereby resulting in further soil acidification.

II. Effects of Acidic Deposition on Soil Invertebrate Communities
A. Responses of Soil Invertebrate Populations to Soil Acidification

The soil-inhabiting invertebrates, which are essential components of the soil decomposer food web and nutrient cycling pathways, are important agents in the formation and maintenance of the biological, chemical, and physical characteristics of the fauna of all soil ecosystems. The soil micro- and mesofauna, which includes the most abundant Nematoda, Acari, and Collembola, as well as macroinvertebrates such as Lumbricidae, Gastropoda, Myriapoda, and some pterygote insects, play key roles in regulating primary decomposer activity, the rates of organic matter degradation, and subsequent release of nutrients.

Because of the complexity of the structure and function of the soil invertebrate community and the great diversity and range of abundance of soil animals, very few studies have considered the effects of acidic deposition on soil ecosystems. Those that did have focused primarily on the effects of acidic deposition on the most abundant faunal groups such as Collembola, Acari, and Enchytraeidae (potworms). Very little information is available on the effects of acidic deposition on Lumbricidae, terrestrial Gastropoda, Isopoda, Myriapoda, Araneae, Protura, Diplura, and most of the soil pterygote insects. Most of the relatively sparse studies of these effects have been based either on artificial addition of simulated acidic deposition or on studies of populations in gradients of natural acidic deposition.

The experimental design of most of the research involves application of solutions of sulfuric acid (occasionally in combination with nitric acid) that are adjusted to different pH levels to soil ecosystems. In other experiments, soil pH has been manipulated by liming. The effects of the exogenous nitrogen and sulfur have usually been studied by application of acidic and neutral sulfate solutions or nitrogenous fertilizers, or by fumigation with SO_2. Most studies have been made on experimental plots in the field or in laboratory microcosms and only for short periods of time. Long-term studies (> 10 yr) of the effects of acidic deposition are few. These include surveys along a point-source pollution gradient (Killham and Wainwright 1981) or a regional acidic deposition gradient (Kuperman 1995). The results of the more important studies are summarized in Table 1.

The most detailed studies of the effects of acidic deposition on soil invertebrate populations have been made in Scandinavian coniferous forests (Abrahamsen 1972a,b, 1983; Abrahamsen et al. 1980; Bääth et al.

Table 1. Summary of laboratory and field experiments testing the effects of acidic and acidifying inputs on soil invertebrate populations. (+, significant increase; 0, no difference; −, significant decrease; and +/−, +/0, 0/−, 0/+, mixed results.)

Animals	Ecological group[a]	Treatment	Dose and form of treatment	Effect	Reference[b]
Ditylenchus spp.	Root/fungal feeders	Liming	9.7 g Ca(OH)$_2$ kg^{-1}	−	23
		NPK (20-6-6)	500 kg ha^{-1}	−	27
		NH$_4$NO$_3$	150 kg N ha^{-1}	0	23
		Urea	150 kg N ha^{-1}	0	23
		Liming	9.7 g Ca(OH)$_2$ kg^{-1}	−	23
		NPK (20-6-6)	500 kg ha^{-1}	0	27
Aphelenchus avenae	Root/fungal feeders	NPK (20-6-6)	500 kg ha^{-1}	−	27
Aphelenchoides spp.	Root/fungal feeders	NH$_4$NO$_3$	150 kg N ha^{-1}	0/−	23
		Urea	150 kg N ha^{-1}	0/−	23
		Liming	9.7 g Ca(OH)$_2$ kg^{-1}	−	23
		NPK (20-6-6)	500 kg ha^{-1}	0	27
Tylencholaimus spp.	Root/fungal feeders	NH$_4$NO$_3$	150 kg N ha^{-1}	+/−	23
		Urea	150 kg N ha^{-1}	+/−	23
		Liming	9.7 g Ca(OH)$_2$ kg^{-1}	0	23
Boleodorus sp.	Root/fungal feeders	NPK (20-6-6)	500 kg ha^{-1}	− −	27

	Feeding group	Treatment	Amount	Effect	Ref.
Acrobeloides nanus	Bacterial feeders	NPK (20-6-6)	500 kg ha^{-1}	+	27
A. emarginatus	Bacterial feeders	NPK (20-6-6)	500 kg ha^{-1}	+	27
Acrobeles sp.	Bacterial feeders	NPK (20-6-6)	500 kg ha^{-1}	−	27
Chiloplacus spp.	Bacterial feeders	NPK (20-6-6)	500 kg ha^{-1}	+	27
Cervidellus sp.	Bacterial feeders	NPK (20-6-6)	500 kg ha^{-1}	−	27
Rhabditis spp.	Bacterial feeders	NH$_4$NO$_3$	150 kg N ha^{-1}	+ / −	23
		Urea	150 kg N ha^{-1}	+ / + +	23
		Liming	9.7 g Ca(OH)$_2$ kg^{-1}	+ +	23
		NPK (20-6-6)	500 kg ha^{-1}	+ +	27
Cephalobus persegnis	Bacterial feeders	NPK (20-6-6)	500 kg ha^{-1}	+	27
Cephalobidae spp.	Bacterial feeders	NH$_4$NO$_3$	150 kg N ha^{-1}	+ / −	23
		Urea	150 kg N ha^{-1}	0 / +	23
		Liming	9.7 g Ca(OH)$_2$ kg^{-1}	− / +	23
		NPK (20-6-6)	500 kg ha^{-1}	+	27
Eucephalobus striatus	Bacterial feeders	NPK (20-6-6)	500 kg ha^{-1}	+	27
E. mucronatus	Bacterial feeders	NPK (20-6-6)	500 kg ha^{-1}	−	27

(continued)

Table 1. (*Continued*)

Animals	Ecological group[a]	Treatment	Dose and form of treatment	Effect	Reference[b]
E. oxyuroides	Bacterial feeders	NPK (20-6-6)	500 kg ha^{-1}	+	27
Teratocephalida spp.	Bacterial feeders	NH$_4$NO$_3$	150 kg N ha^{-1}	−	23
		Urea	150 kg N ha^{-1}	0	23
		Liming	9.7 g Ca(OH)$_2$ kg^{-1}	−/+	23
Plectus spp.	Bacterial feeders	NH$_4$NO$_3$	150 kg N ha^{-1}	−	23
		Urea	150 kg N ha^{-1}	0/−	23
		Liming	9.7 g Ca(OH)$_2$ kg^{-1}	−/+	23
		NPK (20-6-6)	500 kg ha^{-1}	0	27
Wilsomena spp.	Bacterial feeders	NH$_4$NO$_3$	150 kg N ha^{-1}	−	23
		Urea	150 kg N ha^{-1}	−	23
		Liming	9.7 g Ca(OH)$_2$ kg^{-1}	−	23
Monhystera spp.	Bacterial feeders	NH$_4$NO$_3$	150 kg N ha^{-1}	+/−	23
		Urea	150 kg N ha^{-1}	0/−	23
		Liming	9.7 g Ca(OH)$_2$ kg^{-1}	−	23
		NPK (20-6-6)	500 kg ha^{-1}	+	27
Alaimus spp.	Bacterial feeders	NH$_4$NO$_3$	150 kg N ha^{-1}	−	23
		Urea	150 kg N ha^{-1}	−	23
		Liming	9.7 g Ca(OH)$_2$ kg^{-1}	−	23
		NPK (20-6-6)	500 kg ha^{-1}	−	27

Species	Feeding type	Treatment	Dose/pH	Effect	Reference
Steinernema kraussei	Entomophilic	Citric acid phosphate-buffer	pH 3.3–3.6	–	5
			pH 3.7–4.2	–[c]	5
			pH 5.2–6.6	+[c]	5
Clarkus papillatus	Predatory	NPK (20-6-6)	500 kg ha^{-1}	0	27
Mylonchulus brachyuris	Predatory	NPK (20-6-6)	500 kg ha^{-1}	+	27
Enchytraeidae	Ingest plant remains with microorganisms	Urea + PK	200 kg N ha^{-1}	–	1
		Urea	200 kg N ha^{-1}	–	1
		Urea	460 kg N ha^{-1}	–	1
		Ammonium nitrate	200 kg N ha^{-1}	–	1
		Liming	1960 kg Ca ha^{-1}	0	10
		H$_2$SO$_4$	50–150 kg ha^{-1}	0	10
		H$_2$SO$_4$	900 kg ha^{-1}	–	19
		Liming	Crushed CaCO$_3$	–	15
Cognettia sphagnetorum	Ingest plant remains with microorganisms	H$_2$SO$_4$	pH 4.0	+	13
		H$_2$SO$_4$	pH 2.5	–	13, 14, 15
		Liming	3000 kg CaO ha^{-1}	–	13, 14
Mesenchytraeus pelicencis	Ingest plant remains with microorganisms	H$_2$SO$_4$	pH 4.0	+	13
		H$_2$SO$_4$	pH 2.5	–	13
		Liming	3000 kg CaO ha^{-1}	–	13
Enchytronia parva	Ingest plant remains with microorganisms	H$_2$SO$_4$	pH 4.0	+	13
		H$_2$SO$_4$	pH 2.5	–	13
		Liming	3000 kg CaO ha^{-1}	+	13, 14
Lumbricidae, total	Saprophagous	Urea	200 kg N ha^{-1}	0	1
		NH$_4$NO$_3$	200 kg N ha^{-1}	0	1
		H$_2$SO$_4$ + HNO$_3$	pH 3.6 and 4.3	–	12

(continued)

Table 1. (*Continued*)

Animals	Ecological group[a]	Treatment	Dose and form of treatment	Effect	Reference[b]
Eisenia foetida	Epigaeic	12 N HCl	pH <5 and >9	—/—	17
Lumbricus rubellus	Epigaeic	Ammonium sulfate	180 kg N ha^{-1} yr^{-1}	—	18
		S-Coated urea	180 kg N ha^{-1} yr^{-1}	—	18
		Nitrochalk	180 kg N ha^{-1} yr^{-1}	0/—	18
L. castaneus	Epigaeic	Ammonium sulfate	180 kg N ha^{-1} yr^{-1}	—	18
		S-Coated urea	180 kg N ha^{-1} yr^{-1}	—	18
		Nitrochalk	180 kg N ha^{-1} yr^{-1}	0/—	18
Aporrectodea caliginosa	Endogaeic	Ammonium sulfate	180 kg N ha^{-1} yr^{-1}	—	18
		S-Coated urea	180 kg N ha^{-1} yr^{-1}	—	18
		Nitrochalk	180 kg N ha^{-1} yr^{-1}	0/—	18
A. rosea	Endogaeic	Ammonium sulfate	180 kg N ha^{-1} yr^{-1}	—	18
		S-Coated urea	180 kg N ha^{-1} yr^{-1}	—	18

Species	Feeding group	Treatment	Dose	Effect	Ref.
Allobophora chlorotica	Endogaeic	Nitrochalk	180 kg N ha^{-1} yr^{-1}	0/ –	18
		Ammonium sulfate	180 kg N ha^{-1} yr^{-1}	–	18
		S-Coated urea	180 kg N ha^{-1} yr^{-1}	–	18
A. caliginosa caliginosa	Endogaeic	Nitrochalk	180 kg N ha^{-1} yr^{-1}	0/ –	18
		Ammonium sulfate	180 kg N ha^{-1} yr^{-1}	–	18
		S-Coated urea	180 kg N ha^{-1} yr^{-1}	–	18
A. c. tuberculata	Endogaeic	Nitrochalk	180 kg N ha^{-1} yr^{-1}	0/ –	18
		Ammonium sulfate	180 kg N ha^{-1} yr^{-1}	–	18
		S-Coated urea	180 kg N ha^{-1} yr^{-1}	–	18
Dendrobaena rubida (cocoon production)		Nitrochalk	180 kg N ha^{-1} yr^{-1}	0/ –	18
		0.125 *M* H$_2$SO$_4$	pH 4.5	–	21
Gastropoda	Fungivores/ herbivores				
Pinctum pygmaeum	Fungivores/ herbivores	Liming	Crushed dolomite 2, 5, and 10 t ha^{-1}	+ +	28

(continued)

Table 1. (Continued)

Animals	Ecological group[a]	Treatment	Dose and form of treatment	Effect	Reference[b]
Euconulus fulvus	Fungivores/herbivores	Liming	Crushed dolomite 2, 5, and 10 t ha^{-1}	+ +	28
Nesovitrea hammonis	Fungivores/herbivores	Liming	Crushed dolomite 2, 5, and 10 t ha^{-1}	+ +	28
Vertigo substriata	Fungivores/herbivores	Liming	Crushed dolomite 2, 5, and 10 t ha^{-1}	+ +	28
Columella aspera	Fungivores/herbivores	Liming	Crushed dolomite 2, 5, and 10 t ha^{-1}	+	28
Acanthinula aculeata	Fungivores/herbivores	Liming	Crushed dolomite 2, 5, and 10 t ha^{-1}	+ +	28
Discus rotundatus	Fungivores/herbivores	Liming	Crushed dolomite 2, 5, and 10 t ha^{-1}	+	28
Clausilia bidentata	Fungivores/herbivores	Liming	Crushed dolomite 2, 5, and 10 t ha^{-1}	0	28
Oxychilus alliarius	Fungivores/herbivores	Liming	Crushed dolomite 2, 5, and 10 t ha^{-1}	0	28
Vitrina pellucida	Fungivores/herbivores	Liming	Crushed dolomite 2, 5, and 10 t ha^{-1}	0	28
Microarthropoda, total		K$_2$SO$_4$	11.2 mmol m^{-2} SO$_4^{2-}$/application	+ +	16
		KHSO$_4$	11.2–56.2 mmol m^{-2} H^{+}/application	+ +	16

		Treatment	Amount	Effect	Reference
Fungivores		K_2SO_4	11.2 mmol m^{-2} SO_4^{2-}/application	+	16
		$KHSO_4$	11.2–56.2 mmol m^{-2} H$^+$/application	+	16
Predators		K_2SO_4	11.2 mmol m^{-2} SO_4^{2-}/application m^{-2} H$^+$/application	+	16
Detritivores		K_2SO_4	11.2 mmol m^{-2} SO_4^{2-}/application	+	16
Acari, total		Urea	200 kg N ha^{-1}	0	1, 2
		Ammonium nitrate	200 kg N ha^{-1}	0	1
		Liming	Crushed $CaCO_3$	–	6, 8, 14, 15
		Liming	Crushed $CaCO_3$	–	13
		H_2SO_4	pH 2.0–4.0	–/+	6, 8, 11, 14, 15
		H_2SO_4	pH 2.5–4.0	0	13
		H_2SO_4	900 kg ha^{-1}	0	19
Mesostigmata, total		Liming	Crushed $CaCO_3$	–	6, 8, 14
		H_2SO_4	pH 2.0–4.0	–	6, 7, 8, 14
		H_2SO_4	50–150 kg ha^{-1}	–	10
		H_2SO_4 + HNO_3	pH 3.6 and 4.3	0	12
Eviphis ostrinus		H_2SO_4	pH 2.0–5.6	+	7
Pergamasus robustus		H_2SO_4	pH 2.0–3.0	–	7
Saprosecans baloghi	Predator	Urea	200 kg N ha^{-1}	+	1

(continued)

R.G. Kuperman and C.A. Edwards

Table 1. (Continued)

Animals	Ecological group[a]	Treatment	Dose and form of treatment	Effect	Reference[b]
Trachytes sp.		Liming	Crushed CaCO$_3$	+ +	6, 14
		H$_2$SO$_4$	pH 2.0–4.0	– –	6, 7, 14
		H$_2$SO$_4$	50–150 kg ha^{-1}	– –	10
		H$_2$SO$_4$	900 kg ha^{-1}	– –	19
Veigaia nemorensis		Liming	Crushed CaCO$_3$	0	6
		H$_2$SO$_4$	pH 2.0–4.0	0	6, 7
Zerconidae		Liming	Crushed CaCO$_3$	– –	6
		H$_2$SO$_4$	pH 2.0–4.0	– –	6
		H$_2$SO$_4$	50–150 kg ha^{-1}	–	10
Leioseius bicolor, ad.		H$_2$SO$_4$	pH 2.0–4.0	+ +	7
Prozercon kochi		Liming	Crushed CaCO$_3$	0	14
		H$_2$SO$_4$	pH 2.0–4.0	0	7, 14
Parazercon sarekensis		Liming	Crushed CaCO$_3$	– –	14
		H$_2$SO$_4$	pH 2.0–4.0	– –	14
Gamasina	Predator	Liming	Crushed CaCO$_3$	– –	6, 8, 14
		H$_2$SO$_4$	pH 2.0–4.0	– –	6, 7, 8, 13, 14
		H$_2$SO$_4$	50–150 kg ha^{-1}	0	10
Uropodina		Liming	Crushed CaCO$_3$	– –	6
		H$_2$SO$_4$	pH 2.0–4.0	– –	6, 14
		H$_2$SO$_4$	50–150 kg ha^{-1}	–	10
Prostigmata		Urea	460 kg N ha^{-1}	–/+	1, 3

Taxon	Treatment	Condition	Effect	Reference
Oribatida, ad.	Liming	Crushed CaCO$_3$	− −	6
	H$_2$SO$_4$	pH 2.0–4.0	− −	6, 14
	H$_2$SO$_4$	50–150 kg ha^{-1}	−	10
	H$_2$SO$_4$ + HNO$_3$	pH 3.6 and 4.3	+ +	12
	H$_2$SO$_4$	pH 2.0–3.0	+	7
	Urea	460 kg N ha^{-1}	−/+	1, 3
Oribatida, juv.	Urea	460 kg N ha^{-1}	−/+	1, 3
Oribatida, total	Liming	Crushed CaCO$_3$	− −	6, 8, 14
	H$_2$SO$_4$	pH 2.0–4.0	+ +	7, 8, 14
	H$_2$SO$_4$	50–150 kg ha^{-1}	0	10
	H$_2$SO$_4$ + HNO$_3$	pH 3.6 and 4.3	+ +	12
Adoristes ovatus	H$_2$SO$_4$	50–150 kg ha^{-1}	0	10
A. poppei, ad.	H$_2$SO$_4$	pH 2.0–5.3	0	7
Autogneta parva	H$_2$SO$_4$	50–150 kg ha^{-1}	−/+	10
A. tragardhi	H$_2$SO$_4$	50–150 kg ha^{-1}	−/0	10
Belbidae sp.	H$_2$SO$_4$	50–150 kg ha^{-1}	−	10
Brachychochthonius zelawaiensis	H$_2$SO$_4$	pH 2.0–4.0	+/−	7
	Liming	Crushed CaCO$_3$	− −	8
Brachychochthoniidae except *B. zelawaiensis*	H$_2$SO$_4$	pH 2.0–4.0	+ +	7, 14
	Liming	Crushed CaCO$_3$	− −	13
Brachychochthoniidae total	Liming	Crushed CaCO$_3$	− −	8, 13, 14
	H$_2$SO$_4$	pH 2.0–4.0	+ +	8, 13, 14
	H$_2$SO$_4$	pH 2.0–5.3	− −	7
Camisia biurus	H$_2$SO$_4$	50–150 kg ha^{-1}	−	10

(continued)

Table 1. (Continued)

Animals	Ecological group[a]	Treatment	Dose and form of treatment	Effect	Reference[b]
Carabodes spp., ad.		H_2SO_4	50–150 kg ha^{-1}	0/–	10
C. labyrinthicus, ad.		H_2SO_4	pH 2.0–5.3	0	7
Ceratozetes thienemanni, ad.		H_2SO_4	50–150 kg ha^{-1}	–	10
		Liming	Crushed $CaCO_3$	– –	14
Chamobates inscisus		H_2SO_4	50–150 kg ha^{-1}	0	10
Euptyctima, juv.		H_2SO_4	50–150 kg ha^{-1}	+/–	10
Eremeus silvestris		H_2SO_4	50–150 kg ha^{-1}	+	10
Hemileius initialis		H_2SO_4	50–150 kg ha^{-1}	–	10
		H_2SO_4	pH 2.0–5.3	– –	7
Nanhermannia sp.		Liming	Crushed $CaCO_3$	– –	6, 8, 13, 14
		H_2SO_4	50–150 kg ha^{-1}	0	14
Nanhermannia sellnicki, ad.		H_2SO_4	50–150 kg ha^{-1}	+	10
N. sellnicki, juv.		H_2SO_4	50–150 kg ha^{-1}	0/+	10
Northus silvestris		Mixed spruce humus + H_2SO_4 (monoculture)	pH 4.05–4.27	– –	4
		Mixed spruce humus + H_2SO_4 (full fauna)	pH 4.05–4.27	– –	4
		Liming	Crushed $CaCO_3$	– –	6, 8, 13, 14
		H_2SO_4	pH 4.0–2.0	+	7

Species	Treatment	Dose	Effect	Reference
Nothroidea spp., juv.	H₂SO₄	50–150 kg ha⁻¹	0	10
Oppia obsoleta, ad.	H₂SO₄	50–150 kg ha⁻¹	–	10
	H₂SO₄	900 kg ha⁻¹	–	19
	H₂SO₄	pH 2.0–5.3	+/0	7
O. nova, ad.	H₂SO₄	50–150 kg ha⁻¹	+	10, 14
	Liming	Crushed CaCO₃	–	14
	H₂SO₄	pH 2.0–4.0	+/–	7
O. ornuta, ad.	H₂SO₄	pH 2.0–5.3	0	7
Oppioidea spp., juv.	H₂SO₄	50–150 kg ha⁻¹	+/0	10
Paleacarus histricinus	H₂SO₄	50–150 kg ha⁻¹	–	10
Paulonothrus longisetosus, ad. + juv.	H₂SO₄	pH 2.0–5.3	0	7
Phthiracarus spp., ad.	H₂SO₄	50–150 kg ha⁻¹	–	10
Porobelba spinosa	H₂SO₄	pH 2.0–5.3	–	7
Steganacarus sp., ad.	H₂SO₄	pH 2.0–5.3	+ +	7
Suctobelba sp.	Liming	Crushed CaCO₃	–	8, 14
	H₂SO₄	pH 2.0–4.0	–	8, 14
Suctobelba sp., ad.	H₂SO₄	50–150 kg ha⁻¹	–	10
	H₂SO₄	pH 2.0–5.3	0	7
Tectocepheus velatus, ad. + juv.	Liming	Crushed CaCO₃	–	6, 8, 14
	H₂SO₄	pH 2.0–4.0	+ +	6, 7, 8, 14
	H₂SO₄	50–150 kg ha⁻¹	+	10
Trhypochthonius cladonicola	H₂SO₄	50–150 kg ha⁻¹	+/–	10

(continued)

Table 1. (Continued)

Animals	Ecological group[a]	Treatment	Dose and form of treatment	Effect	Reference[b]
Astigmata, total		H_2SO_4	50–150 kg ha^{-1}	−	10
		H_2SO_4 + HNO_3	pH 3.6 and 4.3	+ +	12
		H_2SO_4	pH 2.5–6.0	0	14
		H_2SO_4	pH 2.0–3.0	+	7
Schwiebea cf. lebruni		Mixed spruce humus + H_2SO_4 (monoculture)	pH 4.05–4.27	− −	4
		Mixed spruce humus + H_2SO_4 (full fauna)	pH 4.05–4.27	+ +	4
Rhizoglyphus sp.		H_2SO_4	50–150 kg ha^{-1}	−	10
		H_2SO_4	50–150 kg ha^{-1}	−	10
Collembola, total		Urea	460 kg N ha^{-1}	0	1, 2
		Liming	Crushed $CaCO_3$	− −	8, 9, 13, 14
		H_2SO_4	pH 2.0–4.0	− −	8
		Liming	1960 kg Ca ha^{-1}	0	10
		H_2SO_4	50–150 kg ha^{-1}	0	10
		H_2SO_4		+ +	11
		H_2SO_4	pH 2.5–4.0	+ +	13, 14
Mesaphorura yosii		Mixed spruce humus + H_2SO_4 (monoculture)	pH 4.05–4.27	− −	4
		Mixed spruce humus + H_2SO_4 (full fauna)	pH 4.05–4.27	+ +	4

Species	Treatment	Concentration	Effect	References
	Liming	Crushed CaCO$_3$	0	14
	Liming	Crushed CaCO$_3$	−/− −	8, 9, 13
	H$_2$SO$_4$	pH 2.0–6.0	+/+ +	7, 8, 9, 13, 14
Anurida pygmaea	Liming	Crushed CaCO$_3$	−	8, 9
	H$_2$SO$_4$	50–150 kg ha^{-1}	+/0	10
	Liming	Crushed CaCO$_3$	0	14
A. forsslundi	H$_2$SO$_4$	pH 2.0–4.0	0	7, 14
Anurophorus binoculatus	H$_2$SO$_4$	pH 2.0–5.3	0	7
A. septentrionalis	H$_2$SO$_4$	50–150 kg ha^{-1}	−/0	10
	H$_2$SO$_4$	50–150 kg ha^{-1}	0	10
	Liming	Crushed CaCO$_3$	−	8, 9
Willemia anophthalma	Liming	Crushed CaCO$_3$	− −	8, 9, 13, 14
	H$_2$SO$_4$	pH 2.5–6.0	+ +	8, 9, 13
	H$_2$SO$_4$	pH 2.5–6.0	0	14
	H$_2$SO$_4$	50–150 kg ha^{-1}	−	10
W. aspinata	H$_2$SO$_4$	50–150 kg ha^{-1}	+/0	10
Karlstejnia norvegica	H$_2$SO$_4$	pH 2.5–6.0	+ +	8, 9
Folsomia sensibilis	Liming	Crushed CaCO$_3$	− −	8, 9
F. liisteri	H$_2$SO$_4$	50–150 kg ha^{-1}	0/−	10
Friesea mirabilis	H$_2$SO$_4$	pH 2.0–5.3	0/+	7
Isotomiella minor	Mixed spruce humus + H$_2$SO$_4$ (monoculture)	pH 4.05–4.27	− −	4

(continued)

Table 1. (Continued)

Animals	Ecological group[a]	Treatment	Dose and form of treatment	Effect	Reference[b]
		Mixed spruce humus + H_2SO_4 (full fauna)	pH 4.05–4.27	– –	4, 8
		H_2SO_4	50–150 kg ha^{-1}	–	10
		H_2SO_4	pH 2.0–6.0	0	9, 13
		H_2SO_4	pH 2.0–3.0	–	7
Isotoma notabilis		H_2SO_4	pH 2.0–4.0	– –	7, 8, 9, 13
		H_2SO_4	pH 4.0–6.0	+ +	9
		H_2SO_4	50–150 kg ha^{-1}	–	10
Lepidocyrtus cyaneus		H_2SO_4	pH 2.0–4.0	– –	7, 8, 9
Neelus minimus		H_2SO_4	pH 2.0–5.3	– –	7
Onychiurus absoloni		H_2SO_4	pH 2.0–4.0	–/– –	7, 8
		H_2SO_4	50–150 kg ha^{-1}	+/0	10
O. armatus		H_2SO_4	pH 2.0–4.0	– –	7, 8, 9
		Liming	Crushed $CaCO_3$	– –	8, 9
		H_2SO_4	50–150 kg ha^{-1}	0	10, 14
		H_2SO_4	pH 2.5–6.0	0	14
		H_2SO_4	pH 2.0–6.0	– –	9
Tullbergia krausbaueri		H_2SO_4	50–150 kg ha^{-1}	+ +	10, 13
		H_2SO_4	900 kg ha^{-1}	+ +	19
Xenylla borneri		H_2SO_4	50–150 kg ha^{-1}	–	10
Protura		Liming	Crushed $CaCO_3$	+ +	9

Taxon	Note	Chemical	Dose	Effect	n
Macroarthropoda Total		K_2SO_4	56.2 mmol m^{-2} SO$_4^{2-}$/application	−	16
Litter fungivores		K_2SO_4	56.2 mmol m^{-2} SO$_4^{2-}$/application	−	16
		$KHSO_4$	11.2–56.2 mmol m^{-2} H$^+$/application	−	16
Diplopoda		H_2SO_4	pH 3.0 and 4.0	−	12
Chilopoda		H_2SO_4	pH 3.0 and 4.0	0	12
Coleoptera ad.		Urea	200 kg N ha^{-1}	+	1
		NH_4NO_3	200 kg N ha^{-1}	+	1
Coleoptera lar.		Urea	200 kg N ha^{-1}	+	1
		NH_4NO_3	200 kg N ha^{-1}	−	1
Staphylinidae ad.	Most are predaceous	Urea	200 kg N ha^{-1}	+	1
		NH_4NO_3	200 kg N ha^{-1}	0	1
Staphylinidae lar.	Most are predaceous	Urea	200 kg N ha^{-1}	+	1
		NH_4NO_3	200 kg N ha^{-1}	0	1
Ptiliidae ad.	Fungivores	Urea	200 kg N ha^{-1}	+	1
		NH_4NO_3	200 kg N ha^{-1}	+	1
Cantharidae lar.	Predators	Urea	200 kg N ha^{-1}	0	1
		NH_4NO_3	200 kg N ha^{-1}	0	1
Elateridae lar.		Urea	200 kg N ha^{-1}	−	1
		NH_4NO_3	200 kg N ha^{-1}	−	1
Diptera lar.	Many are saprophagous	Urea	200 kg N ha^{-1}	0	1
		NH_4NO_3	200 kg N ha^{-1}	+	1

(continued)

Table 1. (*Continued*)

Animals	Ecological group[a]	Treatment	Dose and form of treatment	Effect	Reference[b]
Araneae	Predaceous	Urea	200 kg N ha^{-1}	–	1
		NH$_4$NO$_3$	200 kg N ha^{-1}	0	1

[a] ad., Adult; juv., juvenile; lar., larva.

[a] Sources: Baker and Warton 1952; Borror et al. 1976; Hartenstein 1962; Huther 1959; McBrayer et al. 1970; Raw 1967; Wallwork 1967; Wallwork 1970; Watson et al. 1976; Faber 1991; Faber and Verhoef 1991.

[b] 1. Huhta et al. 1986; 2. Vilkamaa and Huhta 1986; 3. Koskenniemi and Huhta 1986; 4. Hågvar 1990; 5. Fischer and Führer 1990; 6. Hågvar and Amundsen 1981; 7. Hågvar and Kjøndal 1981b; 8. Hågvar 1987a; 9. Hågvar 1984a; 10. Bååth et al. 1980; 11. Hågvar and Abrahamsen 1977a; 12. Esher et al. 1992; 13. Abrahamsen et al. 1980; 14. Hågvar 1980; 15. Hågvar 1987b; 16. Craft and Webb 1984; 17. Kaplan et al. 1980; 18. Ma et al. 1990; 19. Lohm 1980; 20. Leetham et al. 1980; 21. Gunnarsson and Rundgren 1986; 22. Marshall 1974; 23. Hyvonen and Huhta 1989; 24. Sohlenius and Wasilewska 1984; 25. Bååth et al. 1978; 26. Berger et al. 1986; 27. Sohlenius and Bostrom 1986; 28. Gärdenfors 1992.

[c] Nematode's ability to parasitize *Cephalcia* nymphs.

1980; Hågvar 1978, 1980, 1984a,b, 1987a,b, 1988a; Hågvar and Abrahamsen 1977a,b, 1984; Hågvar and Amundsen 1981; Hågvar and Kjøndal 1981a,b; Huhta 1984; Huhta et al. 1983, 1986; Koskenniemi and Huhta 1986; Vilkamaa and Huhta 1986). These investigations indicated that acidic deposition can produce either negative or positive changes in populations of different groups of soil invertebrates (Abrahamsen 1983; Bääth et al. 1980; Hågvar 1984a,b; Hågvar and Ammundsen 1981; Heungens and van Daele 1984; Huhta 1984). Hågvar (1984a) suggested that, in general, anthropogenically induced acidification of soils leads to changes in soil invertebrate communities in favor of species that are normally abundant or dominant in naturally acidic soils.

The overall conclusions from these and other studies were that the reactions of soil animals to acidic deposition can be classified into four main categories: (1) an increase in abundance resulting from acidification or a decrease in abundance in response to liming; (2) a decrease in abundance from acidification or an increase in abundance in response to liming; or (3) a decrease in abundance from both acidification and liming (Hågvar 1987a). The effects of acidification on populations of most species of soil-inhabiting invertebrates are included in one of the first two categories, but no species increased its abundance in both liming and acidification experiments (Hågvar 1987a). There were more reports of decreases in populations of soil-inhabiting invertebrates in response to acidification. For instance, in a long-term experiment on liming of acidic grassland, the lower the pH the smaller were populations of most groups of soil- and surface-inhabiting invertebrates (Edwards and Lofty 1975).

Nitrogen is a component of acidic deposition that has the potential to affect overall soil invertebrate populations. Several workers have shown that the abundance of soil-inhabiting invertebrates can be changed considerably in response to additions of different amounts and forms of nitrogen (Abrahamsen 1970; Artemjeva and Gatilova 1975; Axelsson et al. 1973; Behan et al. 1978; Berger et al. 1986; Edwards and Lofty 1969, 1975; Edwards et al. 1976; Escritt and Arthur 1948; Escritt and Lidgate 1964; Gerard and Hay 1979; Huhta et al. 1967, 1969, 1983, 1986; Hyvönen and Huhta 1989; Jefferson 1955; Koskenniemi and Huhta 1986; Lohm et al. 1977; Ma et al. 1990; Marshall 1977; Potter et al. 1985; Rodale 1948; Sohlenius and Bostrom 1986; Sohlenius and Wasilewska 1984; Standen 1984; Tischler 1955; Vilkamaa and Huhta 1986; Zajonc 1975).

Edwards and Lofty (1975) reported that applications of sodium nitrate and ammonium sulfate at rates ranging from 48 to 144 kg ha^{-1} tended to have an overall deleterious effect on total invertebrate populations, but in general they concluded that most forms of nitrogen tended to increase populations in the long term. The effects of acidic deposition on different taxonomic groups of soil-inhabiting invertebrates differed considerably, although there were relatively few data on some groups.

Nematodes.   The effects of soil acidification on the nematode community may be very complex (Dmowska 1995; Heungens 1981; Huhta et al. 1986; Hyvönen and Persson 1990). Several species appeared to be sensitive to decreases in soil pH after application of sulfuric acid. Bacterial-feeding species declined in abundance following artificial acidification (Hyvönen and Persson 1990; Ruess and Funke 1992), while liming increased abundance of the bacterial feeders *Acrobeloides, Protorhabditis,* and *Eumonhystera* and decreased relative abundance of the hyphal feeders (de Goede and Dekker 1993). These results are in agreement with findings of a comparative study of the forest floor invertebrate communities in the northeastern U.S. (Blair et al. 1994). This field survey showed that abundance of fungivores was higher in the hemlock stand with low pH, while bacterivores were relatively more abundant in the red pine stand with higher pH.

Populations of omnivorous, predatory, and saprophytic nematodes were also negatively affected by soil acidification (Dmowska 1993, 1995; Hyvönen and Huhta 1989). Liming increased numbers of omnivores 4 yr after treatment application (de Goede and Dekker 1993). In contrast, plant parasitic nematodes increased their abundance in acidified soils (Brzeski and Dowe 1969; Dmowska 1995). Aphelenchids were insensitive to experimental acidification (Hyvönen and Persson 1990; Ruess and Funke 1992).

The abundance and species richness of nematodes were adversely affected by SO_2 treatment applied at 585 μg m^{-3} (Steiner 1995). The number of species declined in this treatment after 3-6 mon, and only the species *Plectus acuminatus* survived this level of fumigation. Lower SO_2 concentrations (65 and 195 μg m^{-3}) did not affect the species richness of nematodes, but the species *Chiloplectus* cf. *andrassyi* was significantly reduced in population density with increasing SO_2 concentration. *Plectus* cf. *muscorum* and *Prionchulus muscorum* were most frequent in the intermediate treatment levels. In contrast, Leetham et al. (1982) showed that both styletbearing nematodes and Mononchida appeared unaffected by fumigation with SO_2.

Experimental applications of nitrogen (urea) have usually resulted in an increase in numbers of nematodes immediately after treatment (Bassus 1960, 1967; Franz 1959; Heungens 1981; Marshall 1974). Huhta and coworkers (1986) and Hyvönen and Huhta (1989) reported that fertilizing fields with urea increased the abundance of nematodes greatly during the first 2 mon after application, mainly due to increased reproduction of bacterial-feeding species. Similar results were reported by Bääth et al. (1978). In their laboratory experiments, the abundance of the bacterial-feeding nematodes *Rhabditis* and *Diplogaster* increased after the application of ammonium nitrate. An increase in numbers of bacterial-feeding nematodes after fertilization with nitrogen–phosphorus–potassium (NPK) (120 kg N ha^{-1}) was reported by Sohlenius and Bostrom (1986). However, Sohlenius and Wasilewska (1984) reported a decrease in overall numbers of nematodes after the application of ammonium nitrate in the field. Similarly, application of ammonium sulfate decreased the overall abundance of nematodes within 3 wk (Berger et al. 1986), although, 3 mon after a second application, the number of nematodes was 30% greater in the fertilized plots than in the control plots. There is progressively accumulating evidence that nitrogenous fertilizers tend to increase populations of plant-feeding nematodes relative to predatory or detritivorous species (Bohlen and Edwards 1994).

Molluscs. Soil acidification was shown to have a detrimental effect on both abundance and numbers of species of terrestrial molluscs (Valovirta 1968; Waldén 1981; Wärenborn 1982, 1992). This change in abundance is usually attributed to depressed reproduction resulting from the decrease in availability of calcium in acidified soils (Wärenborn 1979). Wärenborn (1992) compared the results of snail fauna surveys conducted approximately 20 yr apart (from mid-1960s to late 1980s) on the same locations in southern Sweden. His study showed that the number of molluscs decreased by 60% in Ca-rich forest sites and by 80% in Ca-impoverished forest sites, and that these decreases were negatively correlated with base saturation. The author concluded that continued acidification and base cation leaching may lead to extinction of the snail fauna in forests of the region.

Earthworms. Very acidic conditions have been reported to decrease the abundance of many species of lumbricid earthworms (Lumbricidae), which tend to occur in low numbers in most naturally acidic forest soils (Blair et al. 1994; Edwards and Bohlen 1996; Edwards and Lofty 1975; Esher et al. 1992; Gates 1978; Hågvar 1980; Kuperman 1996; Ma et al. 1990; Reynolds 1971; Theenhaus and Schaefer 1995). Edwards and Lofty (1975) reported that not all species of earthworms responded to changes in soil pH in the same way, but none of the species they studied could tolerate a pH below 4.0. They reported that individuals of *Lumbricus terrestris* became progressively more numerous as the soil pH increased from 4.0 to 7.5, whereas those of *Allolobophora nocturna, Aporrectodea caliginosa*, and *Aporrectodea rosea* tended to have an optimal pH range of 5.0–6.0, decreasing markedly in numbers at either higher or lower pH levels. A few species of earthworms are relatively tolerant of soil acidity (Edwards and Bohlen 1996). Satchell (1955) reported the lower pH tolerance limit for *Allophorba nocturna, Aporrectodea caliginosa*, and *Aporrectodea rosea* as 4.6 and that for *L. terrestris* and *Octolasium cyaneum* as 4.1, whereas Edwards and Lofty (1975) found that the lower tolerable limit for most species was 4.2.

Multiple application of sulfuric acid with pH 2.7–2.8 led to a significant decrease of the earthworm populations in a spruce forest in southern Germany (Ammer and Makeschin 1994). The species *Lumbricus rubellus* and *Dendrodrilus rubidus* disappeared entirely in this field study. A single experimental field treatment with sulfuric acid (pH 3.0) decreased overall earthworm numbers significantly under a pine plantation (Esher et al. 1993). However, it was impossible to determine whether the earthworms migrated from the acid-treated plots or simply moved deeper into their burrows. Huhta et al. (1986) reported no effects of pH manipulations on earthworm populations in coniferous forest soils in Finland, but the number of earthworms in their study sites was low. In laboratory experiments that tested the effects of acidification on *Eisenia fetida*, 100% mortality occurred at pH below 5.0 or above 9.0 (Kaplan et al. 1980). The abundance of earthworms decreased significantly in oak–hickory forests that had received large amounts of acidic deposition during the past several decades in southern Ohio (U.S.) compared with numbers in ecologically similar forests in southern Illinois that had received much smaller amounts of acidic deposition (Kuperman 1996). Clearly, some species of earthworms can survive at pH levels down to pH 4.0, but many are favored by higher pH, and each species seems to have its individual optimal pH range.

The adverse effects of long-term applications of nitrogenous fertilizers to grassland on populations of lumbricid worms was particularly dramatic in one study (Edwards and Lofty 1975). Populations of all species of earthworms on the site were decreased by the nitrogen applications, but the effect on numbers of *Lumbricus terrestris* was much less than on other species. These results were supported by the findings of Ma et al. (1990), who reported that long-term applications of nitrogenous fertilizers to grasslands had a deleterious effect on earthworm populations in the absence of liming. Application of ammonium sulfate and synthetic sulfur-coated urea caused a significant decrease in the numbers and biomass of earthworms, and endogeic species such as *Aporrectodea caliginosa*, *A. rosea*, *Allolobophora chlorotica*, and *A. caliginosa* were affected much more than epigeic species dominated by *Lumbricus rubellus* and *L. castaneus*. Ammonium N was much more harmful to earthworm populations than nitrate N in a long-term study by Edwards and Lofty (1975), which was in accord with the results of other researchers (Escritt and Arthur 1948; Jefferson 1955; Rodale 1948). Clearly, the effects of acidification on earthworms differ considerably with species but are drastic only at pH levels below 4.0.

Enchytraeidae. Populations of most species of Enchytraeidae decreased in numbers due to a lower soil pH in the spruce forests in Norway (Bääth et al. 1980; Hågvar and Abrahamsen 1977a,b; Lundkvist 1977). Lundkvist reported a 90% decrease in the enchytraeid population in an acidified plot. However, Abrahamsen et al. (1980) reported increased abundance of one enchytraeid species *Cognettia sphagnetorum* after slight soil acidification, although this species was almost eliminated from the most acidified study plots. Similar reactions to acidification and liming were demonstrated by populations of *Mesenchytraeus pelicencic*.

In field studies and laboratory experiments, soil-inhabiting invertebrates were allowed to colonize soil or litter with controlled levels of acidity

(Hågvar 1987a; Hågvar and Abrahamsen 1980; Hågvar and Kjøndal 1981b). In these tests, soil pH influenced the overall success of colonization by a variety of invertebrates, including Enchytraeidae. Edwards and Lofty (1975) reported that nitrogen at low doses had little effect on enchytraeid worms but that higher doses decreased their numbers. In another experiment, populations of enchytraeid worms were affected negatively by nitrogen treatments (urea and NH_4NO_3), and their populations did not recover to those of the controls until the fourth year of the experiment (Huhta et al. 1986). The consensus is that acidic precipitation had adverse effects on enchytraeid populations.

Tardigrades. The response of Tardigrada populations to a range of chronic low-level sulfur dioxide exposure in a northern mixed grass prairie in southeastern Montana was investigated by Leetham et al. (1980). The abundance of Tardigrada decreased substantially in plots exposed to the highest dose (183 μg m^{-3} of SO_2) and was less in the intermediate-dose treatments (105 μg m^{-3} of SO_2); however, these reductions were not statistically significant. Adverse effects on populations of tardigrades of SO_2 treatment applied at 585 μg m^{-3} and in the field surveys along an urban–rural gradient in the region of Zurich, Switzerland were reported by Steiner (1994, 1995). Only the tardigrade species *Echiniscus blumi, Macrobiotus areolatus, Macrobiotus hufelandi*, and *Macrobiotus artipharyngis* survived. At lower SO_2 concentrations (65 and 195 μg m^{-3}), both positive and negative responses to fumigation were observed. There is insufficient evidence to substantiate the overall effects of acidic precipitation on tardigrades.

Acarina. Soil pH can affect the abundance of soil-inhabiting mites. In one study, Oribatidae and populations of all Acari preferred more acidic soils. The species *Tectocepheus velatus, Brachychochthonius zelawaiensis, Nothrus silvestris, Oppia nova, Oppia obsoleta*, and *Schwiebea* cf. *nova* were particularly favored by acidity (Hågvar 1987; Hågvar and Abrahamsen 1980). In other experiments, most species of oribatid mites (Oribatidae), particularly species of Brachychochthoniidae (except *B. zelawaiensis*), *Steganacarus* sp., *T. velatus*, and *O. obsoleta*, increased in abundance in response to acidic treatments, whereas liming usually had a negative effect on populations of these mites; however, a significant decrease in abundance of *Oppia obsoleta* occurred in Swedish soils (Hågvar and Abrahamsen 1980). It should be noted that many species of oribatid mites commonly occur in greater numbers in more acidic soils. Populations of most species of mesostigmatid mites, particularly *Pergamasus robustus, Trachytes* sp., and *Parazercon sarekensis*, as well as species belonging to the Gamasina and Uropodina, reacted negatively to acidification. Exceptions to this were *Eviphis ostrinus* and *Leioseius bicolor*, which increased in abundance (Hågvar 1987a). There was no clear response by populations of prostigmatid mites to acidification. Edwards and Lofty (1975) reported large decreases in populations of mesostigmatid mites (Mesostigmata) in grasslands at pH levels below 5.0.

Huhta et al. (1986) found that no species of mites showed any significant changes in populations in response to treatments with nitrogenous fertilizers (applied in quantities of 200 kg N ha^{-1}), with the exception of one species of mesostigmatid mite, *Saprosecans baloghi*, which occurred in high numbers in plots that had been treated with urea. However, this is a predatory mite, which had probably been distributed onto the soil phoretically by flying insects. A dose of fertilizer containing double the recommended dose of urea resulted in significant decreases in mite populations. In particular, the abundance of prostigmatid and oribatid mites was decreased by the urea treatments, although only a few species of mesostigmatid mites changed significantly in numbers. A general decrease in the abundance of microarthropods (including mites) resulted from application of urea to a Quebec black spruce forest (Behan et al. 1978). An initial decrease was followed by a rapid increase in the populations of these animals. Similar immediate "shock effects" on mite populations were noted by Huhta et al. (1967) during fertilization experiments (NPK and lime) in a Norway spruce forest. The abundance of Mesostigmata, Oribatida, Astigmata, and Prostigmata decreased with increasing nitrogenous inputs. Many species of mites seem to be more tolerant to acidification and to survive well in more acidic soils than some other microarthropods.

Myriapods.   Edwards and Lofty (1975) reported that they did not find any direct correlation between populations of myriapods and soil pH, although numbers of Chilopoda and Symphyla were much larger at pH 5.0–6.0 than at higher or lower pH, and none of these animals occurred at a pH below 4.0. Pauropoda tended to be pH tolerant over the range 4.0–7.0 with no consistent trends in populations. Theenhaus and Schaefer (1995) showed that the increase in soil pH resulted in higher abundance of the chilopod Lithobiomorpha. An increase in numbers of the chilopod *Lithobius* due to liming was also described by Schauermann (1985).

Esher et al. (1992) reported that Diplopoda were not affected by acidification down to pH 4.0–3.0. Myriapoda were affected by nitrogen treatments (Edwards and Lofty 1975). Chilopoda were the most sensitive, being eliminated almost completely from plots regularly treated with the higher doses of N. Symphyla and Pauropoda also decreased considerably in numbers, and Diplopoda were affected least. It does not appear that acidification has severe effects on populations of most groups of myriapods.

Collembola.   Populations of springtails (Collembola) can be affected considerably by acidic deposition, although the responses differ among species. In some studies, numbers of most species increased in response to acidification (Abrahamsen et al. 1980; Bääth et al. 1980; Hågvar 1978, 1980, 1984a, 1987a; Hågvar and Abrahamsen 1977a). Although certain species of springtails preferred more acid soils, others were more abundant in alkaline soils. Populations of the Collembola species *Isotoma notabilis, Isotomiella minor, Lepidocyrtus cyaneus, Neelus minimus, Onychiurus absoloni,* and *Onychiurus armatus* all decreased in response to acidification, whereas liming increased numbers of *Isotoma notabilis* and *Isotomiella minor* (Kreutzer 1995). Other species, including *Anurida pygmaea, Mesaphorura yosii, Willemia anophthalma, Karlstejnia norvegica,* and *Tullbergia krausbaueri,* were more numerous in more acidic soils. The response of *Isotoma notabilis* to acidification was consistent with its distribution in Norwegian soils, being less common in the more acidic soils (Hågvar and Abrahamsen 1984) and in the Netherlands (van Straalen et al. 1987, 1988).

Edwards and Lofty (1975) reported that most Collembola species living in the surface layers of soil (Poduridae and Onychiuridae) were much more numerous in a grassland soil with a nearly neutral pH than in more acidic plots and decreased in numbers dramatically at pH 3.0–4.0. By contrast, populations of species that live on or near the soil surface (Entomobryidae, Sminthuridae, and Isotomidae) differed little between plots with pH ranging from 4.0 to 8.0. In general, it seems that acidification does not favor populations of most species of Collembola, and lower pH levels tend to have adverse effects on populations of many species.

Macroarthropods. Mixed results have been reported in studies of the effects of acid deposition on soil macroarthropod populations. Esher et al. (1992) reported that the total numbers of larger species of macroarthropods were greater in experimental plots treated with sulfuric acid. Wireworms and other insects were relatively numerous in plots with pH between 4.0 and 7.0, but populations were affected drastically by lower pH; very acidic soils did not support large numbers of wireworms.

Effects of acidic and nonacidic SO_4^{2-} on soil arthropod populations were investigated by Craft and Webb (1984) in a mixed oak forest in eastern Tennessee. Neutral sulfate was applied as K_2SO_4, and acidic sulfate was applied as $KHSO_4$. Significant effects of both treatments were observed on litter-inhabiting and soil-inhabiting arthropods. Over a 14-mon period, the numbers of litter macroarthropods were on average 19% lower in the high SO_4^{2-} salt (nonacidic) treatment (10 fold the annual sulfate deposition) than in the low-sulfate treatment (twice the rate of annual sulfate deposition) or the control. Populations of many species of fungivorous macroarthropods also responded negatively to H^+ and SO_4^{2-} treatments. Kuperman (1996) reported that long-term acidic deposition in oak–hickory forests can decrease abundance of total soil macroinvertebrates. Among the groups most sensitive to deposition were fly larvae, termites, and predatory beetles. Kuperman (1996) and Loucks and Kuperman (1991) also suggested that long-term acidic deposition can exacerbate the detrimental effect of certain natural environmental stresses, such as drought, on soil macroarthropod populations and communities.

Among macroarthropods, species in three families of Coleoptera responded to additions of nitrogen (Huhta et al. 1986). Application of urea
caused an increase in Staphylinidae and Ptilidae, whereas populations of
Elateridae larvae responded negatively to nitrogenous fertilization.

Effects of pH manipulation on populations of soil macrofauna caused
by liming of an acidic forest soil were reported by Theenhaus and Schaefer
(1995). Their study showed that an increase in soil pH resulted in significant
changes in the soil macroarthropod community. Several functional groups
of fly larvae, including macrohumiphages (those that feed on litter), micro-
humiphages (which feed on microfungi, algae, Testacea, and amorphic
humus), 'surface scrapers' (which feed on microorganisms, pollen, and
humus on leaf surface), and zoophages (which feed on other animals)
significantly increased in abundance after liming. In contrast, numbers
of mycetophagous and necrophagous macroarthropods were not affected. The abundance of several coleopterans was also affected by soil pH.
The average abundance of carabid beetles and Aleocharinae staphylinid
beetles was higher on the limed plots, while no obvious differences between treatments were observed in the numbers of staphylinid *Othius punc-
tulatus*. *Othius myrmecophilus* had a greater abundance in the more acidic
soils.

In summary, results of most studies showed that acidic deposition can
affect the number of different soil invertebrates. However, there was no
uniform response of soil invertebrates to acidic deposition or treatments
simulating the effects of different components of such deposition. Even
closely related species frequently exhibited different responses to similar
treatments, which included manipulations of soil pH and additions of exogenous nitrogen and sulfur. This points to the need for species-level identification when evaluating the effect of pollution on the soil ecosystem. The
analyses of trends in soil animal responses to acidic deposition and the
ability to draw general conclusions are complicated by the fact that different researchers collected animals from different soils. Overall, experiments
with simulated acid rain suggested that acidic deposition may favor species
with already high abundance in naturally acidic soils, while the numbers of
"calciophilic" species may decline.

B. Possible Mechanisms of Acidic Deposition Effects
on Soil Invertebrate Populations

The response of soil-inhabiting invertebrates to acidic deposition could be
caused by either its direct or indirect effects. Direct effects of acidic deposition on soil-inhabiting invertebrate populations include those due to
changes in soil osmotic potential and changes in soil solution pH resulting
from acidification. Indirect effects include response to changes in the abundance of soil microorganisms that are the primary food for soil-grazing
invertebrates; response to changes in the populations of predators or parasites, which regulate the abundance of invertebrates that are microfloral

grazers; the effects of changes in the substrate quality of the organic matter and responses to changes in the microflora affecting those invertebrates that consume them; and competition between different components of the soil fauna for available substrates.

Direct effects. Acidic deposition can affect populations of soil-inhabiting invertebrates directly by changing the chemical properties of their environment. Such changes caused by acidification include altered osmotic potential, loss of important micronutrients from the soil, and decreases in the soil solution pH. The greatest physiological effects of acid deposition could be expected to be primarily on those groups of soil-inhabiting invertebrates that are in direct contact with the soil water, such as Protozoa, Rotifera, small Nematoda, and small Enchytraeidae (Hågvar 1988a). Atmospheric inputs of H^+ to soil solutions can cause displacement of other cations from soil cation-exchange sites, resulting in a higher soil solution osmotic strength, the so-called salt effect (Myrold 1987). Populations of certain species of soil-inhabiting invertebrates can be affected when the numbers of cations in soil for which they have a high demand are reduced because of acidification. According to van Straalen et al. (1988), the oribatid mite *Platynothrus peltifer* has a high body content of manganese compared to that of other microarthropods and, due to its high requirement for this micronutrient, it appears to avoid acid substrates. In support of this, its abundance was much lower at those sites where acidification had led to a decrease in micronutrients in forest soils.

The addition of large amounts of sulfur and nitrogen to soil through acidic deposition may have negative effects on soil-inhabiting invertebrate populations such as those described in experiments on the effects of fertilizers (Artemjeva and Gatilova 1975; Marshall 1977). Kuhnelt (1961) suggested that both the composition and density of invertebrate populations in soil communities decreased with increasing salt concentration. Moreover, additions of large doses of fertilizers may be responsible for a "shock effect," which results in a rapid decrease in invertebrate populations following application of fertilizers. Similar effects have been reported by others (Hryniuk 1966; Huhta et al. 1967).

Decreased soil pH can lower the availability of some soil nutrients, such as phosphorus, and basic cations that can have an adverse effect on populations with a high demand for that particular nutrient. Land snails, for example, require a source of calcium for normal development and shell production (Crowell 1973; Chétail and Krampitz 1982; Voelker 1959; Wäreborn 1979). However, concentrations of calcium, like other base cations, are affected greatly by acidic deposition. A lower pH in precipitation can result in increased leaching of calcium from organic matter and from exchange sites in soil. In addition, calcium in the foliage in Ca-poor sites may also be decreased. Grädenfors (1992) demonstrated that liming of a relatively nutrient-poor beech forest in southern Sweden had a positive effect on the population densities of land snails.

Decreased pH can also increase the soil solution concentrations of potentially toxic elements such as aluminum and heavy trace metals (Myrold 1987). Several studies have shown that lowering the soil pH resulted in an increased accumulation of certain heavy metals (Cd, Zn, and Pb) by earthworms (Beyer et al. 1987; Ma 1982; Ma et al. 1983).

Hutson (1978) indicated that the pH of the substrate may affect the reproductive success of four species of Collembola. These microarthropods were reared on plaster-charcoal cultures adjusted to pH levels of 2.5–7.6. The acidity gradient influenced the longevity and fecundity of the springtails. These results agreed with those from earlier studies by Ashraf (1969) and Maclagan (1932) that showed the numbers of eggs laid by some species of Collembola depended on the level of soil acidity. Correlations between soil pH and the reproductive success of several species were also proposed as a population-influencing mechanism by Hågvar and Abrahamsen (1980) in laboratory colonization experiments. Some studies have indicated that soil pH may influence springtails by affecting their water uptake through the ventral tube. Other soft-bodied animals (e.g., Lumbricidae, Enchytraeidae, some larvae of Diptera) may also be affected directly by decreased soil acidity. The uptake of salt by water-living soil animals can also be affected by low soil pH, since such effects have been demonstrated for freshwater invertebrates.

Indirect Effects.   Atmospheric deposition of nitrogen and sulfur has been attributed to reductions in soil pH, together with increased SO^{2-} concentrations. Increased amounts of nitrogen in soil can affect soil communities indirectly in several ways, including causing changes in soil acidity, which in turn lead to changes in availability of food (Hyvönen and Huhta 1989), and increases in primary plant production, which result in a higher litter production and a higher rate of root exudation (Sohlenius and Bostrom 1986). Low C:N ratios of plant leaf litter caused by increased nitrogen concentrations in soil can increase the nutrient value of litter to soil-inhabiting invertebrates. Changes in the nutrient status of soil due to high nitrogen may also have an effect on the soil microbial community which in turn has implications for the functioning of soil food webs. Many soil-inhabiting invertebrates feed on fungi and various studies have shown that some fungi can tolerate pH as low as 2.0–3.0.

Bääth et al. (1980) reported that numbers of the collembolan *Tullbergia krausbaueri* were correlated positively with amounts of total fungal mycelium in the A_{00}–A_{02} horizon in a Scots pine forest soil. However, Hågvar and Amundsen (1981) did not find any significant differences in the amounts of fluorescein diacetate-active (FDA-active) and total mycelium in acidified and nonacidified plots whereas populations of fungal-feeding invertebrates, such as oribatid mites, were affected by the acidic treatments. They suggested that, even with an unchanged standing crop of fungal hyphae, the observed changes in populations of mites could be related to the fungal flora because the amount of available hyphae depends mainly on the production rate of the mycelium, which is not necessarily proportional to the fungal biomass. Hågvar and Amundsen also considered the possibility that qualitative changes in soil fungal populations might be important for soil-inhabiting invertebrates. Hågvar and Kjøndal (1981a) suggested that intense acidification of leaf litter can reduce the content of polyphenols or make litter more palatable in other ways for soil macrophytophages, thus leading to an increase in their abundance.

Another possible mechanism, which may explain changes in the abundance of many soil fungal- and detritus-feeding invertebrates following soil acidification, may be the changes in numbers of their predators and parasites. A tendency for a decrease in numbers of predatory mesostigmatid mites to occur in acidified plots has been demonstrated (Abrahamsen et al. 1980). Stinner et al. (1987) also reported decreases in populations of major taxonomic groups of predaceous mites (Mesostigmata, Gamasina) at pH 3.0 compared to pH 5.6. However, in several other studies (Bääth et al. 1980; Hågvar 1984a; Hågvar and Abrahamsen 1980; Hågvar and Amundsen 1981), no decreases in abundance of predatory mesostigmatid mites were observed in response to artificial changes in soil pH.

Changes in soil acidity may also affect the rates of parasitism in certain groups of soil-inhabiting invertebrates. Gunnarsson and Rundgren (1986) studied the risks of mortality of the earthworm *Dendrobaena rubida* in acidified soil. Many of the cocoons produced were infested by rhabditid nematodes, and the incidence of infestation was dependent on pH, in addition to other factors (e.g., timing of cocoon production). The highest infestation occurred at pH 5.5 and at low heavy-metal concentrations. At low pH and high heavy-metal concentrations, few of the cocoons were infested. Similar results were reported for an entomophilic nematode *Steinernema kraussei*, which is an endoparasite of an herbivorous insect, *Cephalcia abietis* (Hymenoptera). Under field and laboratory conditions, positive correlations between soil pH and nematode populations and the numbers of parasitized insects were reported (Fischer and Führer 1990). Soil pH levels below 4.0 have caused limited success in host-finding by this nematode.

Effects on competitive interactions between soil-inhabiting organisms caused by acidic deposition are perhaps the most important mechanism regulating the abundance of different members of the soil invertebrate community in response to acidification and also the most difficult to study. Hågvar (1990) showed that population size (or population growth) of a soil-inhabiting species may be a simple function of soil acidity but can also be modified by other factors, such as the presence of other species, which in turn can be influenced by acidification. He suggested that certain acidophilic microarthropods demonstrate a preference for high pH through increased population growth at higher pH levels. However, this acidophilic characteristic appears only to occur in the presence of whole communities of soil organisms, which may mean that invertebrates can compete with other species more successfully at lower rather than at higher pH. Thus, Håvgar concluded that competition may be a key population-regulating factor of soil-inhabiting invertebrates in response to soil acidification.

III. Effects of Acidic Deposition on Soil Microorganisms

Many soil microorganisms, including viruses, bacteria, actinomycetes, fungi, algae, and protozoa, are closely associated with the soil organic matter. Together with plant roots and invertebrates, these microorganisms are the major components of the overall biomass and biological activity in soils because of their great abundance and their role in the transformation and breakdown of organic matter in terrestrial ecosystems. They are essential catalytic agents of many biochemical reactions in soils and serve as both a source and sink of nutrients contained in organic matter. Soil microorganisms, interacting with soil-inhabiting invertebrates, are responsible for recycling most of the nitrogen, phosphorus, sulfur, and other nutrient elements that are bound in organic matter. They are essential for the formation and maintenance of soil physical structure, the reduction in accumulation of organic matter, and the degradation of many toxic substances introduced into soil.

The effects of acidic deposition on populations of soil microorganisms, and the processes associated with their activities in soil, have long been of considerable interest to microbiologists and other researchers in soil ecology. Many studies have characterized populations of microorganisms in naturally acidic environments, as well as changes in the soil microbial communities resulting from soil acidification by atmospheric deposition. A range of these effects are summarized in Table 2. Unfortunately, many of the earlier studies were made under artificial laboratory conditions, using isolated organisms in pure culture. Although such studies have generated a large volume of basic information, they are of limited use to field ecologists. More recently, more complex laboratory experiments and field studies considering the effects of numerous biotic and abiotic factors have provided much better information on the effects of acidic deposition on populations of different soil microorganisms. There have been several reviews (Alexander 1980a,b; Lettl 1984; Myrold 1990; Myrold and Nason 1992), so we shall summarize some of the earlier findings and focus more on results of the most recent research.

A. Mechanisms of Acidic Deposition Effects on Soil Microorganisms

Most soil microorganisms need an optimal combination of environmental parameters, including physical and chemical characteristics of soil, and availability of nutrients for their successful functioning in soil ecosystems. Increased loading of soils by hydrogen ions, nitrogen, and sulfur by acidic deposition may alter this environmental optimum considerably and have

Table 2. Summary of laboratory and field experiments testing the effects of acidic and acidifying inputs on soil microorganisms. (+, Significant increase, 0, no difference; −, significant decrease; +/−, +/0, 0/−, 0/+, mixed results.)

Soil microorganisms	Type of experiment	Treatment	Dose and form of treatment	Effect	References[a]
Microbial abundance					
Organic horizons	Field	H_2SO_4	50–150 kg ha^{-1} yr^{-1}	−	7, 8
	Field	SO_2 exposure	125 μg m^{-3}	0	10, 46
	Field	SO_2 gradient	300–5000 m	−	12
	Field	SO_2 gradient		−	13, 14
	Laboratory	Lime Ca(OH)$_2$	0–12.2 mg g^{-1}	+	15
Mineral soil	Field	SO_2 gradient	2.8, 6.0, 9.6 km	0	5
Microbial biomass	Field	SO_2 gradient	2.8, 6.0, 9.6 km	−	4, 5
Microbial biomass C	Field	Sulfur	22–44 kg ha^{-1} yr^{-1}	−	1
	Field	$SO_2 + NO_x$ gradient	1.5–57.0 km	0	40
	Chamber trial	$H_2SO_4 + HNO_3$	pH 3.0, 4.0, 4.5, 5.6	0	40, 42
	Field	S gradient	50–750 m	−	43
Bacterial numbers	Chamber trial	SO_2 fumigation	1 mg SO_2 m^{-3}	−	2, 3
	Field	H_2SO_4	50–150 kg ha^{-1} yr^{-1}	−	7, 8
	Field	SO_2 exposure	125 μg m^{-3}	0	10, 46
	Field	SO_2 gradient		−	13, 14
	Field	S gradient	1 m and 200 m	−	11
	Field	$H_2SO_4 + HNO_3$	pH 3.0, 4.2, vs. 5.6	−/0	38
	Field	Magnesian limestone	0–7.5 t ha^{-1}	+	39

(continued)

Table 2. (Continued)

Soil microorganisms	Type of experiment	Treatment	Dose and form of treatment	Effect	References[a]
	Greenhouse	$H_2SO_4 + HNO_3$	pH 4.9, 4.2, 3.5, 2.8	+	48
FDA active	Field	H_2SO_4	50–150 kg ha^{-1} yr^{-1}	–	6
Pseudomonad	Field	$H_2SO_4 + HNO_3$	pH 3.0, 4.2 vs. 5.6	0	38
Cell size	Field	H_2SO_4	50–150 kg ha^{-1} yr^{-1}	–	7, 8
Gram-negative	Greenhouse	$H_2SO_4 + HNO_3$	pH 4.9, 4.2, 3.5, 2.8	+	48
Spore-forming bacteria	Field	SO_2 gradient	2.8, 6.0, 9.6 km	+	4
Starch-utilizing bacteria	Field	SO_2 gradient	2.8, 6.0, 9.6 km	–	4
Sulfate-reducing bacteria	Chamber trial	SO_2 fumigation	1 mg SO_2 m^{-3}	–	2, 3
S-oxidizers	Chamber trial	SO_2 fumigation	1 mg SO_2 m^{-3}	+	2, 3
	Field	SO_2 exposure	125 μg m^{-3}	0	10, 46
Thiosulfate oxidizers	Field	$H_2SO_4 + HNO_3$	pH 3.0, 4.2 vs. 5.6	+/–	38
Nitrifiers	Field	Sulfur	22–44 kg ha^{-1} yr^{-1}	–	1
	Chamber trial	SO_2 fumigation	1 mg SO_2 m^{-3}	–	2, 3
	Field	SO_2 gradient		–	14

Nitrite-oxidizing	Field	$H_2SO_4 + HNO_3$	pH 3.0, 4.2 vs. 5.6	−	38
Ammonium-oxidizing	Field	$H_2SO_4 + HNO_3$	pH 3.0, 4.2 vs. 5.6	0/+	38
Denitrifiers	Laboratory	Lime $Ca(OH)_2$	0–12.2 mg g^{-1}	+	15
Phosphatase-positive	Greenhouse	$H_2SO_4 + HNO_3$	pH 4.9, 4.2, 3.5, 2.8	+	48
Amylolytic	Greenhouse	$H_2SO_4 + HNO_3$	pH 4.9, 4.2, 3.5, 2.8	+	48
Lipolytic	Greenhouse	$H_2SO_4 + HNO_3$	pH 4.9, 4.2, 3.5, 2.8	+	48
Nitrobacter	Laboratory	Lime $Ca(OH)_2$	0–12.2 mg g^{-1}	+	15
N-fixing (free-living)	Field	$H_2SO_4 + HNO_3$	pH 3.0, 4.2 vs. 5.6	−	38
Actinomycetes	Chamber trial	SO_2 fumigation	1 mg SO_2 m^{-3}	−	2, 3
	Field	SO_2 gradient	300–5000 m	−	12
	Greenhouse	$H_2SO_4 + HNO_3$	pH 4.9, 4.2, 3.5, 2.8	+	48
Saprotrophic fungi	Field	SO_2 gradient	2.8, 6.0, 9.6 km	0	4
	Field	SO_2 exposure	125 μg m^{-3}	0	10, 46
	Field	$H_2SO_4 + HNO_3$	pH 3.0, 4.2 vs. 5.6	−/+	38
	Greenhouse	$H_2SO_4 + HNO_3$	pH 4.9, 4.2, 3.5, 2.8	+	48
Total mycelium	Field	H_2SO_4	50–150 kg ha^{-1} yr^{-1}	+	8

(continued)

Table 2. (Continued)

Soil microorganisms	Type of experiment	Treatment	Dose and form of treatment	Effect	References[a]
	Laboratory	Lime Ca(OH)$_2$	0–12.2 mg g^{-1}	–	15
	Field	Magnesian limestone	0–7.5 t ha^{-1}	0	39
	Field	SO$_2$ + NO$_x$ gradient	1.5–57.0 km	–	40
	Laboratory	H$_2$SO$_4$	0.8 kmol H$^+$	0	44
	Laboratory	H$_2$SO$_4$	3.0 kmol H$^+$	0	44
	Laboratory	CaCO$_3$	3000 kg CaCO$_3$ ha^{-1}	0	44
FDA-active	Field	H$_2$SO$_4$	50–150 kg ha^{-1} yr^{-1}	–	6, 7, 8, 9
	Laboratory	H$_2$SO$_4$	0.8 kmol H$^+$	–	44
	Laboratory	H$_2$SO$_4$	3.0 kmol H$^+$	–	44
	Laboratory	CaCO$_3$	3000 kg CaCO$_3$ ha^{-1}	–	44
Penicillium spinulosum	Field	N(NH$_4$NO$_3$)	600 kg N ha^{-1}	+	32
	Field	Urea	600 kg N ha^{-1}	–	32
	Field	H$_2$SO$_4$	50–150 kg ha^{-1} yr^{-1}	+	9
Penicillium cf. *brevi-compactum*	Field	N (NH$_4$NO$_3$)	600 kg N ha^{-1}	–	32
	Field	Urea	600 kg N ha^{-1}	+	32
	Field	H$_2$SO$_4$	50–150 kg ha^{-1} yr^{-1}	0/–	9
P. montanense	Field	H$_2$SO$_4$	50–150 kg ha^{-1} yr^{-1}	0	9
P. brefeldianum	Laboratory	H$_2$SO$_4$	pH 3.5	–	45
Aspergillus niger	Laboratory	H$_2$SO$_4$	pH 3.5	–	45
A. flavipes	Laboratory	H$_2$SO$_4$	pH 3.5	–	45

Species	Type	Treatment	Amount	Effect	Ref
Oidiodendron echinulatum	Field	N (NH$_4$NO$_3$)	600 kg N ha^{-1}	+	32
	Field	Urea	600 kg N ha^{-1}	−	32
	Field	H$_2$SO$_4$	50–150 kg ha^{-1} yr^{-1}	+	9
O. griseum	Field	N (NH$_4$NO$_3$)	600 kg N ha^{-1}	−	32
	Field	Urea	600 kg N ha^{-1}	−	32
O. tenuissimum	Field	H$_2$SO$_4$	50–150 kg ha^{-1} yr^{-1}	0	9
Mortierella sp.	Field	N (NH$_4$NO$_3$)	600 kg N ha^{-1}	−	32
M. isabellina	Field	H$_2$SO$_4$	50–150 kg ha^{-1} yr^{-1}	0	9
M. alliacea	Field	H$_2$SO$_4$	50–150 kg ha^{-1} yr^{-1}	0	9
M. vinacea	Field	H$_2$SO$_4$	50–150 kg ha^{-1} yr^{-1}	0	9
M. macrocystis	Field	H$_2$SO$_4$	50–150 kg ha^{-1} yr^{-1}	0	9
M. ramanniana	Field	H$_2$SO$_4$	50–150 kg ha^{-1} yr^{-1}	0	9
M. parvispora	Field	H$_2$SO$_4$	50–150 kg ha^{-1} yr^{-1}	+/−	9
M. verticillata	Field	H$_2$SO$_4$	50–150 kg ha^{-1} yr^{-1}	0	9
Chaunopycnis alba	Field	N (NH$_4$NO$_3$)	600 kg N ha^{-1}	−	32
	Field	Urea	600 kg N ha^{-1}	−	32
Sagenomella alba	Field	H$_2$SO$_4$	100–150 kg ha^{-1} yr^{-1}	0/+	9
Trichoderma virida	Field	H$_2$SO$_4$	50–150 kg ha^{-1} yr^{-1}	0/+	9
	Laboratory	H$_2$SO$_4$	pH 3.5	−	45
Trichosporiella cerebriformis	Field	H$_2$SO$_4$	50–150 kg ha^{-1} yr^{-1}	0/−	9
Humicolopsis cephalosporioides	Field	H$_2$SO$_4$	50–150 kg ha^{-1} yr^{-1}	0	9
Tolypocladium inflatum	Field	H$_2$SO$_4$	50–150 kg ha^{-1} yr^{-1}	0/−	9

(continued)

R.G. Kuperman and C.A. Edwards

Table 2. (Continued)

Soil microorganisms	Type of experiment	Treatment	Dose and form of treatment	Effect	References[a]
Clitocybe dicolor	Field	N (NH$_4$NO$_3$)	260 kg N ha^{-1}	+	49
			790 kg N ha^{-1}	+	49
C. gibba	Field	N (NH$_4$NO3)	260 kg N ha^{-1}	+	49
			790 kg N ha^{-1}	+	49
C. phyllophila	Field	N (NH$_4$NO$_3$)	260 kg N ha^{-1}	+	49
			790 kg N ha^{-1}	+	49
Lepista inversa	Field	N (NH$_4$NO$_3$)	260 kg N ha^{-1}	+	49
			790 kg N ha^{-1}	+	49
Mycena pura	Field	N (NH$_4$NO$_3$)	260 kg N ha^{-1}	+	49
			790 kg N ha^{-1}	+	49
M. pelianthina	Field	N (NH$_4$NO$_3$)	260 kg N ha^{-1}	+	49
			790 kg N ha^{-1}	+	49
Agaricus abruptibulbus	Field	N (NH$_4$NO$_3$)	260 kg N ha^{-1}	+	49
			790 kg N ha^{-1}	+	49
Lycoperdon perlatum	Field	N (NH$_4$NO$_3$)	260 kg N ha^{-1}	+	49
			790 kg N ha^{-1}	+	49
Collybia butyracea	Field	N (NH$_4$NO$_3$)	260 kg N ha^{-1}	0	49
			790 kg N ha^{-1}	0	49
Clitocybe clavipes	Field	N (NH$_4$NO$_3$)	260 kg N ha^{-1}	0	49
			790 kg N ha^{-1}	0	49

Organism	Test	Treatment	Dose	Effect	Ref.
Wood decomposers	Field	N (NH$_4$NO$_3$)	260 kg N ha^{-1}	0	49
	Field	N (NH$_4$NO$_3$)	790 kg N ha^{-1}	0	49
Mycorrhizal fungi	Field	N (NH$_4$NO$_3$)	260 kg N ha^{-1}	−	49
	Field	N (NH$_4$NO$_3$)	790 kg N ha^{-1}	−	49
Lactarius subdulcis	Field	N (NH$_4$NO$_3$)	260 kg N ha^{-1}	−	49
	Field	N (NH$_4$NO$_3$)	790 kg N ha^{-1}	−	49
Xerocomus chrysentheron	Field	N (NH$_4$NO$_3$)	260 kg N ha^{-1}	−	49
	Field	N (NH$_4$NO$_3$)	790 kg N ha^{-1}	−	49
Ectomycorrhizae	Field	SO$_2$ exposure	0–14.5 ppm hr^{-1}	−	17
	Chamber trial	H$_2$SO$_4$ + HNO$_3$	pH 3.8, 5.2	0	20
	Chamber trial	H$_2$SO$_4$ + HNO$_3$	pH 3.8, 5.2	+	20
	Chamber trial	O$_3$ + SO$_2$ fumigation + acid irrigation	50–180 μg m^{-3} pH 5.6, 4.0	−	21
	Field	H$_2$SO$_4$ + HNO$_3$	pH 4.3, 3.6	−	19
	Laboratory	SO$_4^{2-}$ + NO$_3^-$, 2:1	pH 5.0, 4.0, 3.0	−	17, 25
	Field	Liming	6 mg ha^{-1} season	0	35
Pisolithus tinctorius	Greenhouse	H$_2$SO$_4$ + HNO$_3$	pH 4.8, 4.2, 3.6	+	16
	Laboratory	H$_2$SO$_4$ pH 3.5	Ambient rate	+	22
	Laboratory	H$_2$SO$_4$ pH 3.5	Ambient rate × 3	−	22
Paxillus involutus	Greenhouse	(NH$_4$)$_2$SO$_4$, NaNO$_3$	10–400 kg N ha^{-1} yr^{-1}	−	18
Pisolithus tinctorius	Field	H$_2$SO$_4$	pH 5.6–2.5	−	24
Pisolithus tinctorius strain IBU 1	In vitro	Al$_2$(SO$_4$)$_3$·18H$_2$O	1–20 mM Al	−	36
Pisolithus tinctorius strain IBU 4	In vitro	Al$_2$(SO$_4$)·18H$_2$O	1–20 mM Al	0	36

(continued)

Table 2. (Continued)

Soil microorganisms	Type of experiment	Treatment	Dose and form of treatment	Effect	References[a]
Laccaria bicolor	Field	H_2SO_4	pH 5.6–2.5	0	24
	Laboratory	N (NH₄Cl)	10–20 mg L⁻¹	+	33
	Laboratory	N (NH₄Cl)	200 mg L⁻¹	–	33, 34
	Greenhouse	N (NH₄)₂SO₄	556–778 ppm N	0	47
Laccaria laccata	Laboratory	1.5 $\underline{N}$ H₂SO₄ + Al(SO₄)₃	pH 3.0, 4.0, 5.0 0, 50, 100 mg L⁻¹	–	23
Suillus bovinus	Greenhouse	(NH₄)₂SO₄, NaNO₃	10–400 kg N ha⁻¹ yr⁻¹	–	18
	Laboratory	N (NH₄Cl)	10–20 mg L⁻¹	+	33
	Laboratory	N (NH₄Cl)	100 mg L⁻¹	–	33
Suillus luteus	In vitro	Al₂(SO₄)₃·18H₂O	2 m$\underline{M}$ Al	–	36
	In vitro	Al₂(SO₄)₃·18H₂O	<2 m$\underline{M}$ Al	+	36
Cenococcum geophillim	Laboratory	1.5 $\underline{N}$ H₂SO₄ + Al (SO₄)₃	pH 3.0, 4.0, 5.0 0, 50, 100 mg L⁻¹	–	23
Hebeloma crustuliniforme	Laboratory	1.5 $\underline{N}$ H₂SO₄ + Al (SO₄)₃	pH 3.0, 4.0, 5.0 0, 50, 100 mg L⁻¹	–	23
	Laboratory	NH₄Cl	10–20 mg L⁻¹ N	+	33
	Laboratory	NH₄Cl	200 mg L⁻¹ N	–	33
Amanita muscaria	In vitro	Al₂ (SO₄)₃·18H₂O	2–20 m$\underline{M}$ Al	–	36
	In vitro	Al₂ (SO₄)₃·18H₂O	<2 m$\underline{M}$ Al	0	36
Armillariella mellea	In vitro	Al₂ (SO₄)₃·18H₂O	10–20 m$\underline{M}$ Al	–	36
	In vitro	Al₂ (SO₄)₃·18H₂O	1 m$\underline{M}$ Al	0/+	36

Organism	Setting	Treatment	Dose	Effect	Ref.
Vesicular-arbuscular mycorrhizae (VAM)	Field	SO_2 gradient	5, 13, 20 km	–	26
	Field	SO_2 and NO_x	Regional gradient	0	27
	Field	$(NH_4)_2SO_4$	1.4–4.1 m$\underline{M}$ L^{-1}	0	28
Glomus caledonium	Laboratory	Thiosulfates	12.3–0.12 mg S L^{-1}	+	29
		Metabisulfite	12.3–0.12 mg S L^{-1}	+	29
		Sulfates	12.3–0.12 mg S L^{-1}	+	29
Gigaspora calospora (Nicol. and Gerd.)	Growth chamber	NH_4NO_3	7, 6, 11 mg kg^{-1} N dry soil weekly over 10 wk	+	30
Glomus fasiculatum	Fumigation chamber	SO_2 fumigation of a host (*Phleum pratense*)	0.04–0.065 μL L^{-1} for 6 wk	–	31
Glomus macrocarpum					
Glomus microcarpum					
Glomus mosseae					
Gigaspora sp.					
Protozoa	Field	Sulfur	22–44 kg ha^{-1} yr^{-1}	–	1
Algae	Field	Sulfur	22–44 kg ha^{-1} yr^{-1}	–	1
	Laboratory	1.0 $\underline{N}$ HCl	pH 3.5, 50–300 cm	–	37
	Laboratory	1.0 $\underline{N}$ KOH	pH 5.6, 50–300 cm	+	37

[a]1. Gupta et al. 1988; 2. Mai and Fiedler 1989; 3. Mai and Fiedler 1990; 4. Bewley and Parkinson 1985; 5. Bewley and Parkinson 1984; 6. Bääth et al. 1979; 7. Bääth et al. 1980; 8. Lohm 1980; 9. Bääth et al. 1984; 10. Wainwright 1980; 11. Bryant et al. 1979; 12. Mrkva and Grunda 1969; 13. Langkramer and Lettl 1982; 14. Lettl 1984; 15. Nodar et al. 1992; 16. Walker and McLaughlin 1991; 17. Reich et al. 1985; 18. Termorshuizen and Ket 1991; 19. Esher et al. 1992; 20. Edwards and Kelly 1992; 21. Blaschke 1990; 22. Stroo and Alexander 1985; 23. Entry et al. 1987; 24. McAfee and Fortin 1987; 25. Reich et al. 1986; 26. Clapperton and Parkinson 1990; Clapperton et al. 1990; 27. Heijne et al. 1989a; 28. Heijne et al. 1989b; 29. Hepper 1984; 30. Furlan and Bernier-Cardou 1989; 31. Clapperton and Reid 1992; 32. Arnebrant et al. 1990; 33. Wallander and Nylund 1992; 34. Wallander and Nylund 1991; 35. Antibus and Linkins 1992; 36. Žel et al. 1992; 37. Chang and Alexander 1983; 38. Miller et al. 1991; 39. Shah et al. 1990; 40. Fritze et al. 1992; 41. McColl and Firestone 1987; 42. McColl and Firestone 1991; 43. Visser and Parkinson 1989; 44. Persson et al. 1989; 45. Bewley and Stotzky 1983a,b; 46. Wainwright 1979; 47. Gagnon et al. 1991; 48. Shafer 1992; 49. Rühling and Tyler 1991.

either negative or positive effects on populations of soil microorganisms. The mechanisms of these effects on microbial activity can be either direct or indirect. Direct mechanisms include acidification of the soil solution and increased concentration of anionic components of strong acids (sulfates and nitrates) in soils. The most obvious indirect mechanisms include changes in the quality and availability of organic matter, mobilization of toxic metals in soil, and alteration of symbiotic associations, particularly plant–microbial interactions.

Most soil bacteria are *neutrophiles*, with a pH optimum at or near neutrality and with a range for optimal growth between pH 5 and 8. For many soil fungi, these limits are about one to two pH units lower than for soil bacteria (Myrold and Nason 1992). Although the extracellular concentrations of H^+ that are in direct contact with the cytoplasmic membrane of bacterial cells may vary by nine orders of magnitude (Padan 1984), the intracellular cytoplasmic pH homeostasis must be maintained at levels consistent with the functioning of cellular enzymes. The mechanisms that sustain this homeostasis in the bacterial cell include modified cell membrane permeability, cytoplasmic buffering, acid or base production in the cytoplasm, and active H^+ transport (Booth 1985; Killham et al. 1983; Myrold and Nason 1992; Padan 1984). Many of the same mechanisms are also used by fungi to maintain their cytoplasmic pH homeostasis. Fungal pH optima lower than the bacterial pH optima have been attributed in part to differences in the cell wall structure of fungi and bacteria (Myrold and Nason 1992) and to their ability to sequester excess hydrogen ions in their vacuoles (Padan 1984) in addition to extruding it into the environment.

It is important to distinguish between toxicity of acidic deposition resulting from increased concentrations of H^+ in the soil solution and toxicity of various soluble and oxidation products of gaseous SO_2. Depending on the acidity of the recipient environment, sulfur dioxide can react with water to yield sulfurous acid (H_2SO_3), which may dissociate to yield protons, bisulfite (HSO_3^-), and sulfite (SO_3^{2-}) ions. The sequence of toxicity to soil microorganisms of these ions is $H_2SO_3 > HSO_3^- > SO_3^{2-}$ (Babich and Stotzky 1980). Large amounts of sulfurous acid are present only in the more extremely acidic environments (pH < 2.0) and are of only limited concern. In soils with pH from 1.8 to 7.2, bisulfite is usually the dominant radical; it seems to be the form of SO_2 that exerts the greatest detrimental impact on populations of soil microorganisms (Babich and Stotzky 1978). The HSO_3^- ion reacts with the aldehyde and ketone structures of five and six carbon sugars, with the disulfide linkages of proteins, and with enzyme cofactors such as nicotinamide adenine dinucleotide (NAD^+), flavin mononucleotide, (FMN), and flavin adenine dinucleotide (FAD). These cofactors can deaminate cytosine derivatives to uracil compounds, deactivate mRNA, inhibit ion transport across cell membranes, and alter DNA (Babich and Stotzky 1978). However, because H_2SO_4 is a strong acid, the dominant ions in acidic deposition are H^+ and SO_4^{2-}, and it seems that the hydrogen ions are responsible for the toxicity of deposition to soil microorganisms, whereas the sulfate ion is generally considered to be a nutrient.

In addition to sulfur dioxide, NO_x (and its solubility products HNO_2, HNO_3, NO_2^-, and NO_3^-) may also have detrimental effects on soil microorganisms. The most toxic form, resulting from the dissolution of gaseous NO_2 in water, is anionic NO_2^- (nitrite), which is highly bacteriocidal (Babich and Stotzky 1980). This ion has been shown to have mutagenic effects on bacteriophages, plants, and animals (Babich and Stotzky 1980), and low levels can inhibit photosynthesis in blue-green algae (Wodzinski et al. 1978). High concentrations of inorganic nitrogen in soil inhibit the nodulation of legumes by rhizobia and depress rates of nitrogen fixation (Streeter 1989). Blackmer et al. (1980) demonstrated that concentrations of nitrate greater than 50 μg NO_3–N g^{-1} soil could decrease the microbial reduction of nitrous oxide (N_2O) to molecular nitrogen (N_2). However, nitrogen is an essential nutrient for all heterotrophic microorganisms, and most are not strongly affected by NO_3^- inputs.

Changes in soil pH can also affect the form and chemical mobility of metals. The solubility of potentially toxic metals may increase as soil pH decreases (vanLoon 1984; Xian and Shokohifard 1989). Effects of heavy metals on populations of soil microorganisms have been reviewed by Babich and Stotzky (1980). Few experimental studies have addressed the effects of aluminum (Al) on soil microorganisms, and the mechanisms of its toxicity to these organisms are still not fully understood. Illmer et al. (1995) reported decreases in microbial biomass with increasing aluminum concentrations independent of the soil pH level. Munns and Keyser (1981) showed that the sensitivity of cowpea rhizobia to aluminum was greatest at the time of cell division. Wood and Cooper (1988) reported that *Rhizobium friffolii* was more sensitive to aluminum in its log phase of growth than in the stationary phase. Aluminum may also inhibit bacterial replication by binding directly to DNA (Johnson and Wood 1990), although other modes of action are also possible. Wood (1986) and Wood and Cooper (1988) summarized the effects of aluminum toxicity on rhizobia; they suggested that toxicity could be a direct effect of monomeric Al, an indirect effect of polymeric Al in reducing phosphate concentration, and also a direct effect of polymeric Al.

B. Soil Bacteria and Microbial Biomass

Increased inputs of hydrogen ions, nitrogen, and sulfur to soils associated with acidic deposition are capable of altering the existing environmental optima for microorganisms present in soils and can have either positive or negative effects on their numbers and activity, depending on the ability of the organisms to adapt to these alterations through phenotypic or genetic changes. Consequently, there have been mixed results from experiments that involved studying the effects of artificial acidic deposition or liming on soil microbial populations. Reductions in total bacterial numbers in the H and F layers of a Swedish forest soil exposed to acidic treatments were reported by Bääth et al. (1980) and Lohm (1980). Bacterial populations were decreased by the application of sulfuric acid at an annual rate of 150 kg ha^{-1} to almost one-half the numbers in the control plots and were characterized by the increased presence of spore formers. Differences in bacterial biomass caused by acidification were even more pronounced because microorganisms in the acidified plots were considerably smaller in size than those in the control plots.

Liming of acidic soils often resulted in an increase in bacterial biomass (Bääth and Arnebrant 1994; Shah et al. 1990; Smolander and Mälkönen 1994; Smolander et al. 1994; von Lützov et al. 1992; Zelles et al. 1990). Liming also increased microbial biomass in pot experiments (Badalucco et al. 1992; Carter 1986; Illmer and Schinner 1991) and in field experiments 3 yr after application (Kratz et al. 1991) and 12 yr after application (Priha and Smolander 1994). By contrast, Shafer (1992) reported a three- to eight-fold increase in all microbial populations, including bacteria, in the rhizosphere of sorghum × sudangrass hybrid seedlings exposed to simulated acid rain solutions (adjusted with H_2SO_4 + HNO_3 to pH 4.9, 4.2, 3.5, or 2.8). An increase in numbers of spore-forming bacteria in soils exposed to elevated levels of SO_2 was reported by Prescott et al. (1984). The phenotypic response of bacteria to acidic treatments was both as increased sporulation and decreased cell size.

Bewley and Parkinson (1986a,b) analyzed changes in the microbiota in the organic and mineral horizons of soil in a boreal forest of Alberta, Canada, along the SO_2 emission corridor downwind of a sour gas processing plant. Using the decomposition of glucose and vanillin amendments to soil samples as a method of assessing the sensitivity of soil microorganisms to acidic deposition, they concluded that larger amounts of acidic deposition caused a considerable reduction in total microbial biomass. Fewer bacteria were isolated from both mineral and organic soil samples collected close to the source of emission compared with samples collected from more distant sites. The proportion of bisulfite-tolerant bacteria was much higher in soils exposed to greater amounts of deposition. Earlier studies on the same site (Bewley and Parkinson 1984), based on laboratory dilution plate counts, also indicated a reduction in total numbers of bacteria in the soil H/F horizon due to acidic deposition. Decreased bacterial colonization of the uppermost soil horizons, in areas affected by heavy industrial SO_2 emissions in northwestern Bohemia, was reported by Lettl (1984). Results from other studies along point-source gradients of SO_2 have also indicated that SO_2 can have significant detrimental effects on soil bacterial populations (Bryant et al. 1979; Langkramer and Lettl 1982; Mrkva and Grunda 1969; Prescott et al. 1984).

Additions of exogenous nitrogen with acidic deposition can also affect the soil microbial community. Several studies reported a long-term negative effect of different N treatments on soil microbial biomass or activity (Fauci and Dick 1994; Martikainen et al. 1989, 1990; McAndrew and Malhi 1992; Nohrstedt et al. 1989; Smolander et al. 1994). In contrast, Campbell et al. (1995) observed weak evidence of a stimulatory effect on microbial biomass from frequent N applications. The mechanisms of the effects of N additions on biomass and activity of soil microorganisms are not well understood. Smolander et al. (1994) reported that N application did not decrease soil pH and that decreases in biomass and activity, compared to controls, could not be caused by a long-term increase in soil acidity.

Acidification can also cause significant changes in functional characteristics of bacterial populations. Exposure to greater levels of acidic deposition can result in increased abundance of sulfur-oxidizing bacteria (Gupta et al. 1988; Lettl 1984, 1986; Mai 1990; Mai and Fiedler 1989, 1990; Wainwright 1979), denitrifiers (Davidson et al. 1990; McColl and Firestone 1987), and decreased abundance of nitrifiers (Francis 1982; Lettl 1984; Mai and Fiedler 1989, 1990). However, extremely variable results and conflicting conclusions have been produced by different studies. For example, Francis (1982) reported decreases in the numbers of denitrifiers, whereas the abundance of starch-utilizing bacteria was found to either decrease (Bewley and Parkinson 1986a,b; Shafer 1988), increase (Bääth et al. 1980), or remain unchanged (Mancinelli 1986). Shafer (1992) reported that the numbers of different types of microorganisms (gram-negative bacteria, phosphatase-positive bacteria, amylolytic bacteria, lipolytic bacteria, actinomycetes, etc.), relative to the "total" bacteria, were relatively constant in response to acidification and that no trends relative to treatments (simulated acid rain of pH 4.9–2.8) were evident. An increase in the soil mineral nitrogen content decreased rates of nitrogen fixation (Streeter 1989).

Other researchers have not reported any appreciable effect of acidification on populations of soil microorganisms. Falappi et al. (1994) showed no effect of simulated acid rain (sulfuric and nitric acids, pH 3.0) on aerobic heterotrophic bacteria, free-living N-fixing bacteria, *Mycetes*, and ATP content. Bääth et al. (1979) could find no significant differences in the numbers of bacteria in soil treated with artificial acid rain compared with untreated soil; however, the FDA-active bacteria were reduced in acidified plots. McColl and Firestone (1987, 1991) also reported only slight effects of simulated acid rain treatments of HNO_3 and H_2SO_4 (mole ratio, 3 : 1) and pH 5.6–3.0 on the microbial biomass in A-horizon samples of two forest soils in the Sierra Nevada, California. Wainwright (1980) reported that microbial numbers were unaffected by exposure of soil to atmospheric pollution for 1 yr (SO_2, 125 μg m^{-3}), despite the resultant increase in acidity. There were no major differences in the numbers of microorganisms isolated from mineral soil in the vicinity of the SO_2 emission source compared with that of the remote site (Bewley and Parkinson 1984).

C. Saprophytic Fungi

Unlike populations of soil bacteria, fungal populations may be more tolerant of acidic deposition that may reach either the organic or mineral soil
horizons (Bääth et al. 1979, 1980, 1984; Bewley and Parkinson 1984; Lettl
1981a; Wainwright 1979, 1980). Nevertheless, several workers have observed changes in the species composition of fungal communities associated
with soil acidification (Bääth et al. 1984; Falappi et al. 1994; Mrkva and
Grunda 1969; Wainwright 1979). Abrahamsen et al. (1980) analyzed pH
optima for the growth of several species of fungi isolated from needles of
lodgepole pine and Norway spruce. Among fungi that were affected least
by low pH was *Trichoderma harzianum*, whereas the most tolerant basidiomycete was *Collybia acema*, which grew almost as well at pH 3.0 as at pH
4.5. The other basidiomycete present, *Micromphalia pergorans*, had a
higher pH optimum. The fungal species composition was altered by treatments with sulfuric acid applied at an annual rate of 100 and 150 kg ha^{-1}
for 6 yr (Bääth et al. 1984). For instance, *Penicillium spinulosum* and
Oidiodendron cf. *echinulatum* II were more abundant in soil after an increased rate of acid application. Several workers found overall increases in
total amounts of fungal mycelium and in the relative abundance of fungi in
both the organic and mineral horizons of artificially acidified plots compared to controls (Alexander 1980a; Bääth et al. 1980; Lohm 1980). Alexander (1980a) attributed these increases to a lack of competition from soil
bacteria at the lower pH. In contrast, Falappi et al. (1994) reported a
negative effect of simulated acid rain (sulfuric and nitric acids, pH 3.0) on
several fungal species.

Sulfur dioxide has also been reported to have an effect on the occurrence, abundance, and activities of soil fungi (Bewly and Parkinson 1984;
Domanski and Kowalski 1987; Langkramer and Lettl 1982; Wookey et al.
1990). Mrkva and Grunda (1969) observed an increased incidence of species
in the genera *Rhizopus* and *Mucor* in areas affected by high atmospheric
SO_2 pollution. Increases in populations of several species, including *Coniothyrium quercinum* var. *glandicola*, *Cylindrocarpon orthosporum*, and
Penicillium sp., in response to exposure to gaseous SO_2 were reported by
Newsham et al. (1992a). Domanski and Kowalski (1987) reported increases
in the frequency and abundance of a *Coniothyrium* species on Scots pine
needles in response to higher levels of unspecified atmospheric pollutants.
By comparison, sulfur dioxide was toxic to *Cladosporium* species, *Epicoc-
cum nigrum*, *Fusarium* spp., and *Phoma exigua* after field fumigation of
angiosperm tree leaf litters with SO_2 (Newsham et al. 1992a). These results
were consistent with those of McLeod (1988), who also observed decreases
in the extent of *Fusarium* infection in tillers of winter wheat exposed to
treatments of 0.01, 0.025, 0.04, and 0.06 μL^{-1} SO_2 in an open-air fumigation experiment. Marked reductions in numbers of *Cladosporium* species in
response to SO_2 have been recorded on decomposing litter of Norway

spruce by Langkramer and Lettl (1982) and in the phyllosphere of Valencia orange trees by Fenn et al. (1989).

The differences in response to SO_2 treatments reported here suggest that SO_2 is selectively toxic to different species of saprophytic fungi. Newsham et al. (1992a) concluded that SO_2 treatments did not generally have any significant effects on the extent of fungal colonization of several angiosperm tree leaf litters. Instead, fungal species, which decreased in frequency of occurrence in response to SO_2 treatments, appeared to be replaced in the litter by other species that were more tolerant. In contrast, Langkramer and Lettl (1982) found that decreases in populations of several saprophytic fungi, isolated from the soil of Norway spruce stands along the pollution gradient, did not result in a substantial change in the ratios of each species among the five sites. However, it is possible that differences in evironmental factors other than atmospheric SO_2 concentrations among study sites may have produced these results.

Little information exists on the effects of exogenous nitrogen on populations of saprophytic fungi. The increasing levels of mineral nitrogen deposition over Continental and Northern Europe in recent decades have prompted several studies that investigated this issue primarily by assessing the effects of forest fertilization experiments. Several studies reported the effects of large doses (100–300 kg N ha^{-1}) of mineral nitrogen on the sporophore production of macrofungi (Kardell and Eriksson 1987; Rühling and Tyler 1991; Wästerlund 1982). Most of such investigations dealt predominantly with the diversity and abundance of mycorrhizal flora, and information on the effects of fertilization on saprophytic fungi is sparse. There were no reported effects of ammonium nitrate amendments on the decomposing ability of fungi grown in sterile, nitrogen-poor media to decompose organic matter. However, the sporophore production of *Lepista inversa, L. nuda,* and *Clitocybe rivulosa* was favored markedly by application of a NPK Ca fertilizer to a beech stand in France (Garbaye et al. 1979). Application of NPK to a spruce plantation increased the productivity of *Clitocybe dicolor,* but this treatment was reported to have no effect on *Lycoperdon gemmatum* and *Lepista nuda* (Garbaye and Le Tacon 1982). Increased sporophore production by several species of microbial decomposers of leaf litter and humus was also reported by Rühling and Tyler (1991). They studied the effects of 4 yr of ammonium nitrate applications (12 and 36 kg N ha^{-1} on each of 22 occasions) on fruit body production and species richness of the macrofungal flora of a Swedish beech forest. Populations of *Clitocybe* sp., *Lepista inversa, Mycena pura* and *M. pelianthina* were particularly favored by such fertilization. Among important microbial decomposers with no obvious reaction to the ammonium nitrate treatments were *Collybia butyracea* and *Clitocybe clavipes.* Most of the common wood decomposers did not display any obvious response. Similarly, Gorissen et al. (1994) found no effect of ammonium sulfate application for 23 mon at levels of 5, 50, and 200 kg N ha^{-1} yr^{-1} on fungal communities in laboratory experiments.

D. Mycorrhizal Fungi

The positive role of mycorrhizal infections in enhancing plant growth by their symbiotic associations with fine feeder roots and plant physiological processes and nutrient uptake is well established (Clapperton and Reid 1992; Dighton and Mason 1985; Miller 1987). As a result, a considerable amount of attention has been paid to the effects of acidic deposition on populations of these fungi.

Changes in soil chemistry clearly can influence both rates of fungal growth and physiology (Dighton and Jansen 1991). A wide range of responses to acidic deposition by populations of mycorrhizae have been observed. These organisms can be affected either directly by acidic deposition, by changes in soil pH and nutrient status, or indirectly through host responses. Several experiments have indicated that populations of certain species of ectomycorrhizae are not influenced by acidic deposition because they can adapt readily to acidic environments (Hile and Hennen 1969; Lampky and Peterson 1963; Walker 1989). Populations of the micobiont *Laccaria bicolor* were not affected adversely by simulated acid rain of pH 2.5, and populations of *Pisolithus tinctorius* showed the greatest persistence in plots treated with simulated acid rain adjusted to pH 4.0 (McAfee and Fortin 1987). Meier et al. (1989) reported that four morphotypes of ectomycorrhizae observed on red spruce seedlings were not affected by simulated precipitation of pH 5.5, 3.5, or 2.5. Mikola (1973) considered most ectomycorrhizal fungi to be acidophilic and, under experimental conditions, to be capable of good rates of development in culture at low pH.

The tolerance of several species of mycorrhizal fungi to changes in pH between 2.0 and 7.0 pH units was investigated by Hung and Trappe (1983). These authors demonstrated that, when grown in pure culture, different species of mycorrhizal fungi had specific pH optima. Adams and O'Neill (1991) reported that simulated acidic precipitation treatments of pH 3.3 to 5.2 had no effect on the mycorrhizal colonization of loblolly pine roots. Similar results were reported by Edwards and Kelly (1992), who observed no effects of the pH of precipitation on mycorrhizal populations of loblolly pine seedlings, possibly because the pH treatments were not sufficiently acidic (pH 3.8 and 5.2) to elicit responses associated with more extreme treatments. In contrast, Quimet et al. (1995) found that the root colonization by endomycorrhizal fungi was correlated positively with soil pH but correlated negatively with the proportion of H and Al held on the soil exchange complex.

Walker and McLaughlin (1991) examined both individual and interactive effects of simulated acid rain and ectomycorrhizal inoculation on the growth of white oak and loblolly pine. Application of acidic treatments of pH 3.6 stimulated development of *Pisolithus tinctorius* in the inoculated loblolly pine seedlings. Similarly, the degree of infection by *P. tinctorius* was significantly greater in inoculated oaks exposed to simulated acid rain of pH 4.2 than in those treated at pH 4.8. Furthermore, the degree of ectomycorrhizal infection by naturally occurring fungi in uninoculated white oak and loblolly pine, which received treatments of highest acidity, also exceeded that of seedlings exposed to less acidic treatments.

The effect of soil pH on mycorrhizal populations has also been examined by liming of the forest floor. Composition of mycorrhizal populations was strongly affected by liming of coniferous forests (Antibus and Linkins 1992; Erland and Söderström 1991; Lehto 1994). Lehto (1994) reported that liming increased the proportion of dead short root tips and concluded that lime directly and adversely affected mycorrhizae of Norway spruce. The author suggested that both increased ionic strength and increased pH rather than Ca^{2+} were the reasons for this effect. In contrast, liming stimulated and acid irrigation had no effect on the activity of enzymes in the mycorrhizae of the organic layer of a mature Norway spruce stand in southern Germany (Dähne et al. 1995). Antibus and Linkins (1992) reported that liming caused differences in numbers of viable ectomycorrhizae between control and limed plots that were relatively small considering the spatial and seasonal variability. Liming did not appear to affect the diversity of morphological types of ectomycorrhizae, although lime did increase the relative frequency of certain ectomycorrhizal morphotypes. Mixed results of lime application in a Norway spruce stand were reported by Andersson and Söderström (1995). Three types of mycorrhizae decreased after liming, two increased, and one was not consistently affected.

These results contrast with the findings of several other workers who reported that acidic deposition had a detrimental effect on some ectomycorrhizal development. Shafer et al. (1985) reported that application of artificial acid rain of pH 4.0 and 3.6 inhibited the development of two species of ectomycorrhizal fungi, *Thelephora terrestris* and *Laccaria laccata*, on loblolly pine seedlings, compared with the rate of development at pH 5.6 and 2.4. Several other reports indicated that the degree of mycorrhizal infection was affected negatively by artificial acid rain in soil under northern red oaks (Reich et al. 1985, 1986), under white pine seedlings (Reich et al. 1988; Stroo and Alexander 1985), under longleaf pine and under loblolly pine (Esher et al. 1992), under Scots pine seedlings (Dighton and Skeffington 1987), under paper birch seedlings (Keane and Manning 1988), and under Norway spruce (Blaschke 1988, 1990; Göbl 1988). Discrepancies between the results of different experiments that assessed responses of mycorrhizal fungi to acidic deposition can be attributed to any one or a combination of the following: (1) the species-specific sensitivity of organisms to changes in their chemical environment that affect growth, infectivity, and survival directly; (2) the physicochemical characteristics of soils that determine the fate of different components of acidic deposition; (3) the buffering capacity of some soils; (4) differences in host and symbiont species used in the various studies; and (5) differences in experimental design and methods used.

Increased soil loading with nitrogen associated with acidic deposition is another factor that can affect rates of mycorrhizal development. In most natural ecosystems, unaffected by atmospheric pollution, nitrogen is a growth-limiting factor. However, in some parts of the world, anthropogenic nitrogen inputs have reached levels as high as 22 kg N ha^{-1} yr^{-1}, and local extremes may elevate this to 64 kg N ha^{-1} yr^{-1} (Anonymous 1988; van Breemen et al. 1982). These amounts of nitrogen deposition have the potential to affect rates of mycorrhizal development significantly. Several studies examined the effects of N additions to soil in field and pot experiments. Gagnon et al. (1991) reported no effects of nitrogen on ectomycorrhizal formation in containerized red oak seedlings. However, most studies demonstrated that high nitrogen availability can decrease or even inhibit mycorrhizal development (Harley and Smith 1983; Richards 1965; Rühling and Tyler 1991; Tétreault et al. 1978). Almost complete disappearance of the sporophores of 49 mycorrhizal species, in response to applications of ammonium nitrate (12 and 36 kg N per ha on each of 22 occasions) for 4 yr, was reported by Rühling and Tyler (1991). Termorshuizen and Ket (1991) reported that the numbers of mycorrhizae, as well as mycorrhizal frequency, were affected negatively by nitrogen treatments of potted *Pinus sylvestris* seedlings that had established mycorrhizae of *Paxillus involutus* and *Suillus bovinus*. The beginning of this negative effect depended on the availability and form of the mineral nitrogen, the mycorrhizal fungal species, the tree species, and the carbohydrate economy of the plant. The mycorrhizae used in this study were inhibited more by ammonium than by nitrate, indicating the indirect effects of nitrogen via the internal nutritional status of the plant.

Theodorou and Bowen (1969), Bigg (1981), and Rudawska (1986) reported that nitrate nitrogen had a stronger inhibitory effect on mycorrhizal development than ammonium nitrogen. They interpreted this effect as an indirect response to a reduced concentration of carbohydrates in the roots, following an increased conversion of carbohydrates into amino acids, thereby limiting the supply of sugars to mycorrhizae under conditions of high nitrogen availability. Björkman (1942) hypothesized that increasing the absorption of nitrogen by plants could decrease the rate of mycorrhizal development. Bigg (1981) suggested that decreases in carbohydrate concentrations in nitrate-treated roots may result from the energy-consuming reduction of nitrate to ammonium. However, contrary to Björkman's hypotheses, Nylund (1988) and Wallander and Nylund (1991) could not detect any negative correlation between nitrogen supply and carbohydrate concentrations in roots or shoots. Their study showed that variations in shoot nitrogen concentration were enough to explain the variations in ergosterol concentrations (a measure of fungal biomass) better than any combination of factors, including carbohydrate status. These authors interpreted their results as proving that nitrogen had an effect on mycorrhizal formation that is not mediated through its effects on carbohydrate concentrations. They concluded that Björkman's hypothesis regarding the action of nitrogen on populations of mycorrhizae is not tenable. They admitted, however, that their results did not completely contradict the existence of effects of carbohydrate.

A decrease in soil pH from acidic deposition can also cause changes in soil chemistry. Soil solutions in neutral soils usually contain about 400 μg L^{-1} Al, while at pH 4.4 Al reaches 5700 μg L^{-1} (Kabata-Pendias and Pendias 1986). The resultant increased solubilization of toxic metals, particularly aluminum, may influence different physiological processes such as the structure and function of membranes, mineral metabolism, and water uptake, which in turn, can affect the development and vitality of mycorrhizal fungi (Burt et al. 1986; Dighton and Skeffington 1987; Entry et al. 1987; McQuattie and Schier 1987; Žel and Gogala 1990; Žel et al. 1992). Thompson and Medve (1984) reported a 30%-60% decrease in hyphal growth of *Cencoccum geophilum* Fr., *Pisolithus tinctorius*, and *Thelephora terrestris* by 150 μM Al, whereas the growth of *Suillus luteus* did not decrease significantly until the Al concentrations reached 1050 μM. Growth of *T. terrestris* and *P. tinctorius* decreased at levels of 600 and 2000 μM Mn, whereas the growth of *S. luteus* and *C. geophilum* did not decrease in response to any level of this metal used. Also, different isolates of *P. tinctorius* could react differently to Al treatments. Strain IBU 1 was inhibited by 1-20 mM Al, while there were no significant differences between the growth of strain IBU 4 (Žel et al. 1992) and the control. Increased solubility of Al was shown to have an inhibitory effect on protein synthesis and OH^- and H^+ transport systems in mycorrhizal fungi (Oelbe-Farivar 1985) and on the lateral transport of minerals, especially Ca and P (and possibly also on membrane fluidity), in mycorrhizal fungi (Žel et al. 1992).

Acidic deposition on vesicular-arbuscular mycorrhizae (VAM) has also yielded mixed results. VAM can be affected adversely by acidic deposition and exposure to ozone (Brewer and Heagle 1983; Ho and Trappe 1984; McCool et al. 1979; Rice et al. 1979). Hayman and Travares (1985) suggested that extreme levels of soil pH may limit the ability of VAM to develop the extramatrical hyphae necessary for phosphate uptake from the bulk soil but that this does not limit the infectivity of VAM. A negative correlation of VAM infection in root samples with soil pH was reported by Noordwijk and Hairiah (1986). Chambers et al. (1980) observed decreased VAM development in response to applications of ammonium sulfate. A similar effect resulting from additions of nitrogen was reported by Hayman (1970), although experiments by Hepper (1983) showed that lettuce growing on nutrient solutions with higher amounts of nitrogen had greater infections of VAM than in solutions with less nitrogen. These results agreed with those of Furlan and Bernier-Cardou (1989), who reported a stimulation by nitrogen of onion root colonization by VAM.

Neither negative nor positive effects of ammonium sulfate application on VAM infection in the field could be detected by Heijne et al. (1989a). A significant decrease in the VAM inoculum potential was identified in soils that had been exposed to high SO_2 levels in proximity to a sour gas processing plant, in comparison to the potential in more remote areas (Clapperton and Parkinson 1990). Exposure of *Phleum pratense* L. to SO_2 in controlled-environment fumigation chambers reduced the root lengths colonized by VAM fungi and the lengths of roots infected with arbuscules (Clapperton and Reid 1992). These authors attributed these effects to changes in the availability of photoassimilates for fungal metabolism. Such changes could have resulted from detrimental effects of SO_2 fumigation on phloem loading and the speed at which assimilates are translocated from the leaves (Gould et al. 1988). However, Hepper (1984) reported that higher concentrations of inorganic sulfur could stimulate growth of VAM hyphae from germinating spores. No effect of simulated acid rain at pH 3 and 4 on VAM development was observed by Killham and Firestone (1983). No correlation could be detected between amount of VAM infection and amount of artificial acid rain (Heijne et al. 1989b). Visser et al. (1987) concluded that there is no current evidence to suggest any direct effects of SO_2 on these organisms and that extreme changes in soil pH would be necessary to have any significant effects on VAM. It appears that, like their ectomycorrhizal and saprophytic counterparts, different species of VAM may have different soil pH optima.

In summary, many studies have shown that changes in soil pH and additions of exogenous nitrogen and sulfur with this deposition can affect the successful functioning of microbial communities in soil ecosystems by changing the optimal combination of environmental parameters of soil or by altering plant–microbial interactions of symbiotic associations. Because many soil bacteria are neutrophiles, decreases in soil pH frequently have had adverse effects on their numbers. However, depending on the ability of soil bacteria to adapt to these changes, both positive and negative effects of soil acidification on bacteria have been reported. Soil fungi, having lower pH optima, are generally more tolerant of the effects of acid rain. Nevertheless, changes in species composition of fungal communities as a result of soil acidification have been observed. Overall, it is difficult to interpret discrepancies between the results of different experiments with simulated acid rain in the absence of standardized testing procedures. The development of standardized procedures will alleviate comparison problems arising from species-specific sensitivity of soil organisms to changes in their chemical environment, differences in physico-chemical characteristics of soils, differences in host and symbiont species used, and differences in experimental design and methods used in such studies.

IV. Effects of Acidic Deposition on Soil Biological Processes
A. Organic Matter Breakdown

Organic matter decomposition and the mineralization of nutrients contained within it are critical processes affecting the overall availability of plant nutrients and the primary productivity of ecosystems. Decomposition processes can influence both the structural and functional nature of forest ecosystems. The recycling of the nutrients contained in leaf litter is an important component of ecosystem dynamics, and the regulation of rates and timing of nutrient release play an indispensable role in ecosystem functioning. The bulk of the annual aboveground net primary productivity of an oak–hickory forest ecosystem is transferred directly to the decomposer subsystem as foliar and woody litter. Therefore, an assessment of the rates of litter decomposition and of the rates and synchrony of nutrient retention and release is essential to understanding how acidic deposition can affect ecosystem functioning.

In previous sections, we discussed how different constituents of acidic deposition can influence the soil physical and chemical environment and the soil biota inhabiting it. Hence, the rates of breakdown of organic material produced in the ecosystems and transferred to the decomposer subsystem as foliar and woody litter can also be affected by acidic deposition because of the crucial role of the soil biota in the decomposition process.

Most experiments examining the effects of acidic deposition on organic matter decomposition applied solutions of sulfuric and/or nitric acids to soils. Results of both field and laboratory decomposition studies have shown various effects of simulated acid rain on rates of leaf litter decomposition (Table 3). These effects depended largely on several factors, including the length of exposure of the plant litter to acidic treatment, the level of acidity in the simulated precipitation, the type of experiment (microcosm incubation or litter bag field study), and the species of plant litter used. Results of nearly half of the studies could not demonstrate any statistically significant effects of wet acidic deposition on rates of litter decomposition when treatments were applied at environmentally realistic concentrations and, in some studies, even when the pH of the solution applied was as low as 2.0 (Abrahamsen et al. 1980; Berg 1986a,b; Chang and Alexander 1984; Gray and Ineson 1981; Hägvar and Kjøndal 1981a; Larkin and Kelly 1987, 1988a,b; Lee and Weber 1983; Neuvonen and Suomela 1990; Williams 1988). Few studies have reported any stimulating effect of simulated acid rain on rates of mass loss of plant litter or rates of CO_2 evolution from decomposing plant material (Lee and Weber 1983; Roberts et al. 1980). Effects of increased exogenous nitrogen supply with acidic deposition on litter decomposition differed among different plant substrates. Nitrogen inputs stimulated decay rate of *Carex* species litter (Aerts et al. 1995) and cellulose (Entry and Backman 1995) but did not affect decomposition of jack pine (*Pinus banksiana* Lamb.) needle litter (Prescott 1995).

Table 3. Summary of laboratory and field experiments testing the effects of acidic and acidifying inputs and liming on soil microbial activity. (+, significant increase; 0, no significant difference; −, significant decrease; +/−, +/0, 0/−, 0/+, mixed results.)

Soil ecosystem parameter	Type of experiment	Treatment	Dose and form of treatment	Effect	References[a]
Basal respiration	Field	Sulfur	22–44 kg ha^{-1} yr^{-1}	−	1
	Field	Sulfur	1072–35800 μg g^{-1}	0	41
	Field	SO$_2$ gradient	2.8, 6.0, 9.6 km	−	4, 5, 6, 29
	Field	SO$_2$ gradient	57-km transect	0	38
	Field	SO$_2$ gradient	3.5–34.3 km	−	21
	Chamber trial	SO$_2$ fumigation	1 mg SO$_2$ m^{-3}	−	2, 3
	Field	H$_2$SO$_4$	50–150 kg SO$_2$ ha^{-1} yr^{-1}	−	7, 10
	Field	H$_2$SO$_4$	6 × 150 kg ha^{-1} yr^{-1}	−	43
	Field	H$_2$SO$_4$	pH 2.0, 3.0	−	42
	Field	SO$_2$ exposure	125 μg m^{-3}	0	11
	Field	S gradient	1 m and 200 m	−	12
	Field	SO$_2$ gradient		0	49
	Field	H$_2$SO$_4$, NHO$_3$	0.05–0.5 mol m^{-1}	0	51
	Field	H$_2$SO$_4$ + NHO$_3$	pH 2.0	−	35
	Field	H$_2$SO$_4$ + NHO$_3$	pH 3.0, 4.0	+	35
	Laboratory	H$_2$SO$_4$	pH 3.0, 2.5	0	36
	Laboratory	H$_2$SO$_4$	pH > 3.0	0	37
	Field	Liming	7.5 t ha^{-1}	+	39
	Field	Liming	1.96 t Ca ha^{-1}	+	43
	Field	Liming	0.47 kg rock m^{-2}	0	52
	Field	Liming	2.45 t Ca, 0.14 t Mg, 0.03 t K ha^{-1}	+	59

System	Treatment	Dose	Effect	Reference
Laboratory	H_2SO_4 + NHO_3	pH 3.0, 4.0, 4.5	0	40, 55
Laboratory	Acid rain	pH 3.0, 3.7	0	46
Laboratory		pH 3.0, 4.0, 5.5	0	47
Laboratory	H_2SO_4 + NHO_3	pH 3.5–5.7	0	57
Laboratory		pH 2.4	−	56
Laboratory	Liming	pH 5.0–7.0	+	56
Laboratory	Liming	3000 kg $CaCO_3$ ha^{-1}	+	54
Laboratory	Liming	4000 kg $CaCO_3$ + $MgCO_3$ ha^{-1}	+	58
Laboratory	H_2SO_4	38–150 kg ha^{-1}	−	54
Laboratory	Soil pH 3.2	1 N $\underline{H_2SO_4}$	−	19
Laboratory	Soil pH 7.1	$Ca(OH)_2$	+	19
Substrate-induced respiration (SIR)				
With glucose				
Field	SO_2 gradient	2.8, 6.0, 9.6 km	−	4, 5
Field	SO_2 gradient	2.8, 6.0, 9.6 km	0	6
Field	S gradient	1 m and 200 m	+	12
With vanillin				
Field	SO_2 gradient	2.8, 6.0, 9.6 km	−	4, 5, 6
With cellulose				
Field	SO_2 gradient	2.8, 6.0, 9.6 km	−	6
Field	S gradient	1 m and 200 m	−	12
With starch				
Field	S gradient	1 m and 200 m	−	12
With urea				
Field	SO_2 gradient	2.8, 6.0, 9.6 km	−	6
Field	S gradient	1 m and 200 m	−	12
With casein				
Field	S gradient	1 m and 200 m	0	12
NO$_3$-N content				
Chamber trial	SO_2 fumigation	1 mg SO_2 m^{-3}	−	2, 3
Enzyme activities				
Cellulase				
Field	SO_2 exposure	125 μg m^{-3}	0	11
Laboratory	H_2SO_4	pH 3.0, 2.5	−	36

(continued)

Table 3. (Continued)

Soil ecosystem parameter	Type of experiment	Treatment	Dose and form of treatment	Effect	References[a]
Dehydrogenase	Field	Sulfur	22–44 kg ha⁻¹ yr⁻¹	−	1
	Field	SO₂ exposure	125 µg m⁻³	0	11
	Field	H₂SO₄ + NHO₃	pH 2.0	−	35
	Field	H₂SO₄ + NHO₃	pH 3.0, 4.0	+	35
	Laboratory	H₂SO₄ + NHO₃	pH 3.0, 3.7	0	46
	Field	H₂SO₄ + NHO₃	pH 3.0, 3.7	0	63
	Laboratory	Soil pH 3.2	1 N H₂SO₄	−	19
	Laboratory	Soil pH 7.1	Ca(OH)₂	0	19
	Greenhouse	H₂SO₄	pH 3.3, 4.3	−	60
	Field	SO₂ gradient		−/0	49
Urease	Field	Sulfur	22–44 kg ha⁻¹ yr⁻¹	−	1
	Field	SO₂ exposure	125 µg m⁻³	0	11
	Field	H₂SO₄ + NHO₃	pH 2.0	−	35
	Field	H₂SO₄ + NHO₃	pH 4.0	+	35
	Field	H₂SO₄ + NHO₃	pH 3.0, 3.7	0	63
	Laboratory	Soil pH 3.2	1 N H₂SO₄	−	19
	Laboratory	Soil pH 7.1	Ca(OH)₂	0	19
Phosphatase	Field	SO₂ exposure	125 µg m⁻³	0	11
	Field	H₂SO₄ + NHO₃	pH 2.0	−	35
	Field	H₂SO₄ + NHO₃	pH 3.0, 4.0	0	35
	Laboratory	H₂SO₄ + NHO₃	pH 3.0, 3.7	0	46
	Field	H₂SO₄ + NHO₃	pH 3.0, 3.7	0	63
	Field	SO₂ gradient		0	49

	Study type	Treatment	Dose	Effect	Ref.
	Laboratory	Liming of soil	$Ca(OH)_2$ 0–4.3 mg g^{-1}	–/+	50
	Field	Liming	0.47 kg rock m^{-2}	0	52
	Field	Liming	6 kg Mg ha^{-1} yr^{-1}	–	62
Alkaline phosphatase	Field	Sulfur	22–44 kg ha^{-1} yr^{-1}	–	1
Acid phosphatase	Field	SO$_2$ gradient	3.5–34.3 km	–	21
	Greenhouse	H$_2$SO$_4$	pH 3.3, 4.3	–	60
Arylsulfatase	Field	Sulfur	22–44 kg ha^{-1} yr^{-1}	–	1
	Field	SO$_2$ exposure	>125 μg m^{-3}	0	11
	Field	H$_2$SO$_4$ + NHO$_3$	pH 2.0	–	35
	Field	H$_2$SO$_4$ + NHO$_3$	pH 4.0	0	35
	Field	H$_2$SO$_4$ + NHO$_3$	pH 3.0	+	35
	Greenhouse	H$_2$SO$_4$ + O$_3$ <160 nL O$_3$ L^{-1}	pH 3.3, 4.3	+	60
	Greenhouse	H$_2$SO$_4$ + O$_3$ >160 nL O$_3$ L^{-1}	pH 4.3	–	60
Rhodanese	Field	SO$_2$ exposure	>125 μg m^{-3}	0	11
	Field	SO$_2$ exposure	>125 μg m^{-3}	+	61
Xylanase	Field	Liming	0.47 kg rock m^{-2}	0	52
Protease	Laboratory	H$_2$SO$_4$ + NHO$_3$	pH 3.0, 3.7	0	46
	Field	H$_2$SO$_4$ + NHO$_3$	pH 3.0, 3.7	–	63
	Laboratory	Liming of soil	Ca(OH)$_2$ 0–4.3 mg g^{-1}	+	50
	Field	Liming	0.47 kg rock m^{-2}	+	52
Nitrogenase	Field	SO$_2$ gradient		0	49
Sulfatase	Laboratory	Liming of soil	Ca(OH)$_2$ 0–4.3 mg g^{-1}	+	50
Decomposition Litter: Needle litter	Field	H$_2$SO$_4$	50–150 kg ha^{-1} yr^{-1}	–	8, 9

(continued)

Table 3. (Continued)

Soil ecosystem parameter	Type of experiment	Treatment	Dose and form of treatment	Effect	References[a]
	Field	Liming	Ca, 1960 kg ha⁻¹	−	8
	Field	H_2SO_4 pH 3.1, 2.7	25 and 50 kg S ha⁻¹ yr⁻¹	0	24
	Field	Urea-N	150 kg N ha⁻¹	−	28
	Field	SO_2 gradient	2.8, 6.0, 9.6 km	−	29
Abies balsamea + Picea rubens 2:1	In vitro	H_2SO_4 + NHO_3	pH 3.0–4.0	−	20
P. abies	Laboratory	H_2SO_4	pH 3.0, 2.0	0	25
P. mariana	Field	H_2SO_4 + NHO_3	pH 3.0, 4.0, 5.5	0	47
Pinus strobus	Field	SO_2 gradient	3.5–34.3 km	−	21
P. sylvestris	Field	H_2SO_4 + NHO_3	pH 3.0, 4.0	0	23
	Field	SO_2 fumigation	≤0.048 μL L⁻¹ for 215 d	−	33
	Field	SO_2 gradient	57-km transect	0	38
P. contorta	Laboratory	H_2SO_4	pH 3.0, 5.7	0	48
		Liming	3000 kg CaO ha⁻¹	−	48
Pinus spp	Field	H_2SO_4	25 and 50 kg	+	31
Leaf litter	Field	S inputs	Long-term (30 yr)	0	30
Betula pubescens + B. verrucosa	Field	H_2SO_4	pH 2.0	−	13
			pH 3.0, 4.0	0	13
Betula pubescens + B. verrucosa	Greenhouse	H_2SO_4	pH 2.0	−	13
			pH 3.0, 4.0	0	13

Species	Type	Treatment	Condition	Response	Ref.
B. papyrifera	Field	SO$_2$ gradient	3.5–34.3 km	—	21
B. tortuosa	Field	H$_2$SO$_4$ + NHO$_3$	pH 3.0, 4.0	—	23
Betula spp.	Field	SO$_2$ fumigation	0.01–0.03 μL L^{-1} SO$_2$ for 16–68 wk	0	15
Acer macrophyllum	Field	H$_2$SO$_4$	pH 3.5	+	14
		H$_2$SO$_4$	pH 3.0, 4.0	—	14
A. pseudoplatanus	Field	SO$_2$ fumigation	0.01–0.03 μL L^{-1} SO$_2$ for 16–68 wk	0	15
Populus tricho-carpa	Field	SO$_2$ gradient	0.5 km vs. non-polluted site	—	22
	Field	Acid rain	H$_2$SO$_4$ pH 3.0–4.0	0	14
P. tremuloides	Field	SO$_2$ gradient	3.5–34.3 km	—	21
Quercus garryana	Field	H$_2$SO$_4$	pH 3.0–4.0	0	14
Quercus palustris	Field	H$_2$SO$_4$	pH 3.0–4.0	+	14
Q. petraea	Field	SO$_2$ fumigation	0.01–0.03 μL L^{-1} SO$_2$ for 16–68 wk	0	15
Quercus spp.	Laboratory	Soil pH 3.2	1 N H$_2$SO$_4$	—	18, 19
		Soil pH 7.1	Ca(OH)$_2$	+	18, 19
Fraxinus latifolia	Field	H$_2$SO$_4$	pH 3.0–4.0	+	14
F. excelsior	Field	SO$_2$ fumigation	0.01–0.03 μL L^{-1} SO$_2$ for 16–68 wk	0	15
Corilus avellana	Field	SO$_2$ fumigation	0.01–0.03 μL L^{-1} SO$_2$ for 16–68 wk	0	15
Ulmmus americana	Field	H$_2$SO$_4$	pH 3.0–4.0	+	14

(continued)

Table 3. (Continued)

Soil ecosystem parameter	Type of experiment	Treatment	Dose and form of treatment	Effect	References[a]
Liriodendron tulipifera	Field	H_2SO_4	pH 3.0–4.0	+	14
Salix scouleriana	Field	H_2SO_4	pH 3.0–4.0	+	14
Liquidambar styraciflua	Field	H_2SO_4	pH 3.0–4.0	0	14
Platanus acerifolia	Field	H_2SO_4	pH 3.0–4.0	0	14
Molinia caerulea	Field	N fertilization		+	27
Agropyron smithii	Field	SO_2 fumigation	$0.076\ \mu L\ L^{-1}$ for 153 d	−	34
Lichen thalli					
Cladina stellaris	Field	$H_2SO_4 + NHO_3$	pH 3.0, 4.0, 5.5	0	47
Root litter	Field	H_2SO_4	$50–150\ kg\ ha^{-1}\ yr^{-1}$	−	8, 9
Humus	Laboratory	H_2SO_4	pH 3.4	−	16
Soil organic matter	Laboratory	H_2SO_4	pH 3.5, 4.1 for 14 d	−	17
Wood					
Thuja plicata	Laboratory	H_2SO_4, NHO_3, H_2SO_3	pH 3.5, 4.0	0	26
		H_2SO_4, NHO_3, H_2SO_3	pH 2.0, 2.5, 3.0	−	26

Calluna vulgaris	Field	N fertilization	pH 2	+	27
Populus tremula	Field	H_2SO_4	4500 kg CaO ha^{-1}	-/0	32
	Field	Liming		+/0	32
Mineralization					
Cellulose	Chamber trial	SO_2 fumigation	1 mg SO_2 m^{-3}	-	2, 3
	Field	SO_2 gradient	2.8, 6.0, 9.6 km	-	6
	Field	SO_2 gradient	0.5 km vs. nonpolluted site	-	22
	Laboratory	H_2SO_4 + NHO_3	pH 3.8	0	20
	Field	N fertilization		+	27
	Field	H_2SO_4	pH 2.0	-/0	32
	Field	Liming	4500 kg CaO ha^{-1}	+/0	32
Glucose	Field	SO_2 gradient	2.8, 6.0, 9.6 km	0	5
	Laboratory	H_2SO_4	pH 2.0	-	37
	Field	Sulfur	1072–35800 μg g^{-1}	-	41
Vanillin	Field	SO_2 gradient	2.8, 6.0, 9.6 km	-	5
Urea	Field	SO_2 gradient	2.8, 6.0, 9.6 km	-	6
Casein	Field	SO_2 gradient	2.8, 6.0, 9.6 km	-	6
N Turnover	Field	H_2SO_4	6 × 150 kg ha^{-1} yr^{-1}	-	43
Net N mineralization	Laboratory	Liming of soil	$Ca(OH)_2$ 0–4.3 mg g^{-1}	+	50
	Field	H_2SO_4, NHO_3	0.05–0.5 mol m^{-1}	0	51
	Laboratory	Liming	3000 kg $CaCO_3$ ha^{-1}	+	54
	Laboratory	H_2SO_4	38–150 kg ha^{-1}	+	54
N fixation	Laboratory	Acid rain	pH 3.5, 5.6 21 d	-	44
	Laboratory	H_2SO_4 + NHO_3	pH 3.0, 3.7	0	46
	Field	H_2SO_4 + NHO_3	pH 3.0 vs. 5.6	-	45

(continued)

R.G. Kuperman and C.A. Edwards

Table 3. (*Continued*)

Soil ecosystem parameter	Type of experiment	Treatment	Dose and form of treatment	Effect	References[a]
Ammonification	Field	SO_2 gradient	2.8, 6.0, 9.6 km	–	6
	Field	SO_2 exposure	125 μg m^{-3}	+	11
	Field	H_2SO_4	6×150 kg ha^{-1} yr^{-1}	0	43
	Laboratory	Soil pH 3.2	1 N H_2SO_4	–	19
		Soil pH 7.1	Ca(OH)$_2$	+	19
Nitrification	Field	SO_2 exposure	125 μg m^{-3}	0	11
	Field	H_2SO_4, NHO$_3$	0.05–0.5 mol m^{-1}	0	51
	Laboratory	H_2SO_4	pH 4.4–4.1	–	48
		Liming	1.5–6 T CaO ha^{-1}	+	48
	Field	Liming	0.47 kg rock m^{-2}	+	52
	Laboratory	H_2SO_4 + NHO$_3$	pH 2.0	–	53
	Laboratory	Soil pH 3.2	1 N H_2SO_4	+	19
		Soil pH 7.1	Ca(OH)$_2$	+	19
	Laboratory	H_2SO_4 + NHO$_3$	pH 3.0	–	55
Denitrification	Laboratory	H_2SO_4 + NHO$_3$	pH 2.0	+	53
	Laboratory	Soil pH 3.2	1 N H_2SO_4	–	19
		Soil pH 7.1	Ca(OH)$_2$	+	19
	Laboratory	H_2SO_4 + NHO$_3$	pH 3.0	+	55

[a] 1. Gupta et al. 1988; 2. Mai and Fiedler 1989; 3. Mai and Fiedler 1990; 4. Bewley and Parkinson 1985; 5. Bewley and Parkinson 1986a,b; 6. Bewley and Parkinson 1984; 7. Bääth et al. 1979; 8. Bääth et al. 1980b; 9. Lohm 1980; 10. Bääth et al. 1984; 11. Wainwright 1980; 12. Bryant et al. 1979; 13. Hågvar and Kjøndal 1981a; 14. Lee and Weber 1983; 15. Newsham et al. 1992c; 16. Hågvar 1988b; 17. Chang and Alexander 1984; 18. Francis 1982; 19. Francis et al. 1980; 20. Moloney et al. 1983; 21. Freedman and Hutchinson 1980; 22. Killham and Wainwright 1981; 23. Neuvonen and Suomela 1990; 24. Gray and Ineson 1981; 25. Hovland et al. 1980; 26. Williams 1988; 27. French 1988; 28. Titus and Malcolm 1987; 29. Prescott and Parkinson 1985; 30. Larkin and Kelly 1987; 31. Roberts et al. 1980; 32. Abrahamsen et al. 1980; 33. Wookey et al. 1990; 34. Dodd and Lauenroth 1981; 35. Killham et al. 1983; 36. Hovland 1981; 37. Bewley and Stotzky 1983a,b; 38. Fritze et al. 1992; 39. Shah et al. 1990; 40. McColl and Firestone 1991; 41. Visser and Parkinson 1989; 42. Bääth et al. 1979; 43. Lohm et al. 1984; 44. Chang and Alexander 1983; 45. Miller et al. 1991; 46. Bitton and Boylan 1985; 47. Moore 1987; 48. Abrahamsen et al. 1976; 49. Nohrstedt 1985; 50. Haynes and Swift 1988; 51. Johnson and Todd 1984; 52. von Mersi et al. 1992; 53. Firestone et al. 1984; 54. Persson et al. 1989; 55. McColl and Firestone 1987; 56. Wilhelmi and Rother 1990; 57. Cronan 1985; 58. Zelles et al. 1987a,b; 59. Marschner and Wilczynski 1991; 60. Reddy et al. 1991; 61. Wainwright 1979; 62. Antibus and Linkins 1992; 63. Bitton et al. 1985.

In studies where application of acidic treatments or placement of litter bags in study plots along an established acidic deposition gradient resulted in reductions in rate of plant litter breakdown, the inhibitory effects of wet acidic deposition were usually attributed to the detrimental effect of this treatment on the microbial activity in decomposing litter (Bääth et al. 1980; Francis 1982; Francis et al. 1980; Freedman and Hutchinson 1980; Hägvar and Kjøndal 1981a; Killham and Wainwright 1981; Lohm 1980; Moloney et al. 1983; Neuvonen and Suomela 1990; Prescott and Parkinson 1985).

However, Newsham et al. (1992b) argued that the use of simulated wet deposition is largely inappropriate in studies of the effects of acidic precipitation on soil and soil processes, because the main pathway by which sulfur and nitrogen oxides are transferred to the ground is through dry deposition (Fowler 1984; Garland 1977; Williams et al. 1989). The presence of cations (calcium, magnesium, ammonium, etc.) in wet precipitation (NAPAP 1990) may neutralize the effects of sulfate in wet-deposited precipitation before it contacts the litter or soil. As a result, wet-deposited sulfur dioxide in the form of sulfate may not exert the same degree of influence on the chemical contents of soil and litter as would an equivalent input of dry-deposited SO_2 (Newsham et al. 1992b). The inhibitory effect of dry-deposited SO_2 in field fumigation experiments on degradation of western wheatgrass (*Agropyron smithii*) and Scots pine (*Pinus silvestris*) litters was demonstrated by Dodd and Lauenroth (1981), Wookey (1988), and Wookey et al. (1990), respectively. Similar effects of dry deposition on rates of degradation of western wheatgrass litter were also reported in laboratory experiments (Leetham et al. 1983). Exposure of Austrian pine (*Pinus nigra*) leaf litter to gaseous SO_2 for 82 d in the laboratory resulted in a 33% reduction in rates of microbial respiration (Ineson 1983).

In addition, Wookey (1988) found statistically significant inverse correlations between leaf litter respiration rates, dry weight losses, and SO_2 concentrations in a long-term study (204 d) using field fumigation experiments with angiosperm tree leaf litters. Similar inverse correlations between rates of litter decomposition and SO_2 concentrations in air were demonstrated by Newsham et al. (1992b,c, 1995). Their experiments reported the toxic effects of SO_2 on populations of members of the decomposer subsystem and marked reductions in amounts of calcium and magnesium in field-fumigated litters. These two essential elements in biological systems could account for the reductions in CO_2 evolution from decaying leaf litters and, consequently, the inhibition of litter decomposition rates. However, Fritze et al. (1992) found no decreases in the rates of green needle litter decomposition from SO_2 pollution. These authors attributed the lack of treatment effect to the fact that they buried litter bags in the humus horizon where they were fairly well buffered against the effects of acidic pollutants, whereas, in other studies (e.g., Berg 1986a; Neuvonen and Suomela 1990), litter bags were placed on top of the litter layer. Fritze et al. concluded that the latter technique should be used in such studies because it approximates the natural situation more closely and places the litter in direct contact with the depositing pollutants.

B. Soil Respiration

There has been considerable research into the effects of acidic deposition on soil respiration (see Table 3). Several field studies have demonstrated no effects of acidic inputs on C mineralization in soils of deciduous (Johnson and Todd 1984; Kelly and Strickland 1984), coniferous (Roberts et al. 1980; Visser and Parkinson 1989), or mixed forests (Bitton et al. 1985) and along an SO_2 gradient (Wainwright 1980). These findings were supported by negative results of laboratory incubations of unamended organic and mineral soils with low to moderate levels of acidic inputs (1–10 kmol H^+ ha^{-1} yr^{-1}) (Bääth et al. 1979; Cronan 1985; Hovland 1981; Lohm et al. 1984; McColl and Firestone 1987, 1991; Moore 1987; Popovic 1984; Will et al. 1986). Ljungholm et al. (1979) also reported no effect of decreases in soil pH on microbial respiration as judged by the recorded heat effects. Variable results with occasional increases in CO_2 evolution at these levels of inputs were reported by Killham and Firestone (1982), Killham et al. (1983), Bitton and Boylan (1985), and Wilhelmi and Rother (1990). Decreases in soil respiration at moderate levels of acidic input have been observed in organic soils and litter layers (Bääth et al. 1980, 1984; Bitton and Boylan 1985; Chang and Alexander 1984; Greszta et al. 1992; Lohm et al. 1984; Salonius 1990).

High levels of acidic inputs (> 10 kmol H^+ ha^{-1} yr^{-1}) have resulted in decreases in rates of CO_2 evolution from organic and mineral soils (Bääth et al. 1979; Bryant et al. 1979; Francis 1982; Francis et al. 1980; Hendrikson 1985; Killham et al. 1983; Klein et al. 1984; Persson et al. 1989; Zelles et al. 1987a), in no effects (Cronan 1985; McColl and Firestone 1987, 1991), or in mixed results (Bitton and Boylan 1985; Chang and Alexander 1984; Hovland et al. 1980; Killham and Firestone 1982). Garden and Davies (1988) reported microbial respiration rates of leaf litter exposed to pH 3.0 and 4.0 were significantly lower than those litter exposed to pH 5.4. Liming has been reported to have a stimulative effect on biological activity in various soil horizons (Lang and Beese 1985; Marschner and Wilczynski 1991; von Mersi et al. 1992; Persson et al. 1989; Shah et al. 1990; Zelles et al. 1987a,b, 1990), a negative effect (Lohm et al. 1984), or no clear effects (Bääth et al. 1980).

Increased levels of sulfur deposition can also affect microbial activity in soil. Tamm et al. (1977) observed that additions of sulfur as elemental S caused a decrease in CO_2 evolution even when pH was not affected. These results agree with those of Grant et al. (1979) and Prescott and Parkinson (1985), who found that continuous exposure to SO_2 suppressed heterotrophic microbial activity in an acid soil. This decrease in microbial activity could have resulted from the slight drop in soil pH caused by the treatments or from increased solubilization of toxic metals generated in acidic soil. Larkin and Kelly (1988a) also reported reductions in soil respiration rates following large sulfur additions (~ 66 kg ha^{-1} yr^{-1}). They theorized that there is a response threshold below which additions of sulfur will have a neutral or possible stimulatory impact but above which negative impacts will begin to occur.

The mechanisms of the effects of sulfur inputs other than acidity effects on microbial activity are not well understood. Ghiorse and Alexander (1976) determined that microorganisms were not involved directly in the removal of sulfur dioxide from the atmosphere but that they may play a role in the conversion of sulfite and bisulfite to sulfate. The detrimental effects of bisulfite accumulation were suggested as possible inhibitors of microbial activity by (Babich and Stotzky (1978) and Bewley and Parkinson (1986b); however, bisulfate is usually rapidly converted to sulfate in soil (Grant et al. 1979).

Several studies have investigated the effects of acidic simulation treatments on the evolution of CO_2 from soils amended with various carbon sources (see Table 3). Respiration in treated soils was generally similar to that in unamended soils. Soil respiration after amendments with glucose was affected negatively by decreases in soil pH below 4.0 (Ljungholm et al. 1979; Lohm et al. 1984), 3.8 (Lohm 1980), or 3.0 (Bewley and Stotzky 1983a,b). Amending soils with more complex carbon sources, such as vanillin (Bewley and Stotzky 1984) and cellulose (Ljungholm et al. 1979; Moloney et al. 1983), also caused lower respiration rates after the soil pH fell below 4.0, whereas there was no effect on respiration at higher pH.

The effects of SO_2 on soil respiration have been investigated in point-source gradient studies. No effects of SO_2 pollution on microbial activity in the forest floor (Fritze et al. 1992; Lettl 1984; Nohrstedt 1985) and mineral soil (Lettl 1984; Wainwright 1980) were shown; however, decreases in CO_2 evolution from the forest floor and mineral soil in polluted sites compared with unpolluted sites were reported by Lettl (1984) and Bryant et al. (1979). Significant reductions in rates of CO_2 production from F/H horizon material collected near the source of SO_2 emissions, both in unamended samples and in samples amended with cellulose or urea, as compared with similarly treated samples from sites more remote from pollution were also reported by Bewley and Parkinson (1986a).

C. Enzymatic Activities in Soil

Soil enzymes play an important role in the cycling of nutrients in soil ecosystems, particularly in mineralization and immobilization dynamics. It is generally accepted that enzyme activity in soil is derived ultimately from microbial, plant, and animal sources (Ladd 1978). The syntheses and activity of enzymes are influenced by chemical characteristics of the soil environment, such as pH and carbon sources. Hence, changes in enzyme activity can be useful indicators of the effects of acidic deposition on overall microbial and biochemical activities in soil.

Research into forms of soil enzyme activity is extensive (Burns 1978), but relatively few studies have considered the effects of acidic deposition on enzymatic activities in soil. Results of several such studies are summarized in Table 3. Most commonly, dehydrogenase activity has been used as an index of net overall microbial activity in soils affected by acidic inputs (Bitton and Boylan 1985; Bitton et al. 1985; Francis et al. 1980; Gupta et al. 1988; Killham et al. 1983; Maccari et al. 1994; Nohrstedt 1985; Reddy et al. 1991; Wainwright 1980). Results closely parallel those obtained for soil respiration: dehydrogenase activity was stimulated at low acidic input rates and depressed at higher input rates, with varied effects at intermediate doses. The effects of acidic deposition on cellulase activity followed the same general pattern as those on dehydrogenases (Hovland 1981; Myrold 1987; Wainwright 1980).

Other soil enzymes differed in their sensitivities to acidic treatments. Urease, phosphatase, and arylsulfatase enzymes catalyze the mineralization of organic N, P, and S to inorganic forms, so that changes in their activities can affect the availability of these nutrients to plants. Killham et al. (1983) reported that an acidic input of pH 4.0 stimulated urease activity but did not significantly affect either phosphatase and arylsulfatase activity in surface soil. A pH of 3.0 stimulated arylsulfatase, did not significantly affect urease, and only slightly inhibited phosphatase activity. Reddy et al. (1991) reported that irrigation of loblolly pine seedlings with simulated acid rain of pH 4.3 and 3.3 significantly affected the rhizosphere activity of dehydrogenase, acid phosphatase, and arylsulfatase. Activities of dehydrogenases and acid phosphatases were inhibited linearly as the pH fell, whereas changes in arylsulfatase activity were dependent on specific combinations of simulated acid rain pH and ozone concentration. Decreased activities of dehydrogenases and acid phosphatases at pH 3.0 compared to 3.6 were also reported by Will et al. (1986). However, Bitton et al. (1985) found that pH levels between 3.0 and 3.7 did not affect phosphatase activity. Similarly, acidic inputs had no effects on the activity of arylsulfatase (Wainwright 1980; Will et al. 1986). The optimum pH for arylsulfatase activity in soil ranges from 5.5 to 6.2 (Speir and Ross 1981; Tabatabai and Bremner 1970) and is not affected by the addition of sulfate (Tabatabai and Bremner 1970).

Low to moderate acidic inputs did not effect or even slightly stimulate of protease activity (Bitton and Boylan 1985; Bitton et al. 1985), although liming increased protease activity (Haynes and Swift 1988; von Mersi et al. 1992). Liming either stimulates (Harrison 1987) or inhibits soil phosphatase activity (Antibus and Linkins 1992; Halstead 1964; Haynes and Swift 1988; von Mersi et al. 1992).

Rates of soil microorganism enzyme production appear to depend directly on pH. Nahas et al. (1982) reported that alkaline phosphatase secretion was stimulated and acid phosphatase secretion was inhibited at pH above 7.4 in *Neurospora crassa* whereas the opposite effect occurred at pH below 5.7. Rates of fungal pectinase production were correlated with pH variations in organisms from several taxa, including *Fusarium* (Persley and Page 1971), *Rhizoctonia* (Lisker et al. 1975), *Penicillium* (Phaff 1947), *Botrytis* (Hancock et al. 1964; Leone and Van den Heuvel 1986), *Aspergillus* (Dean and Timberlake 1989a,b; Tuttobello and Mill 1961), and *Saccharomycopsis* (Fellows and Worgan 1984). Aguilar et al. (1991) also reported that production of endo- and exopectinase by *Aspergillus* sp. CH-Y-1043 was influenced by pH of the medium with maximum production attained at pH 2.5.

Enzyme assays of a range of soils obtained from sites along sulfur dioxide pollution gradients showed no effect of the deposition on enzymes (Nohrstedt 1985; Wainwright 1980), except for a decrease in acid phosphatase activity (Freedman and Hutchinson 1980) and an increase in rhodanase activity (Wainwright 1979). In contrast, Maccari et al. (1994) reported a 20%–40% decrease in phosphatase, β-glucosidase, dehydrogenase, benzoyl-argininamide protease, and casein protease activities in soil exposed to high levels of SO_2 (429 kg ha^{-1} yr^{-1}) in a microcosm study.

D. Nutrient Cycles

The Nitrogen Cycle.   Availability of nitrogen is often a factor for limiting primary productivity in many terrestrial ecosystems and, after carbon, it is the element that exerts the greatest influence on the overall biological activity of soil ecosystems. Movement of nitrogen between the atmosphere, land, and water is a part of global nitrogen cycling, but soil has an internal nitrogen cycle in which organic forms of nitrogen are converted to mineral forms. This conversion results from the biological activity of soil organisms and consists of several distinct steps. During the first step, ammonification, organic N is converted to NH_4^+ by heterotrophic microorganisms. This step is followed by nitrification, during which NH_4^+ is further oxidized to form NO_3^-. Nitrification arises primarily from the combined activities of autotrophic bacteria, including *Nitrosomonas*, which converts NH_4^+ to NO_2^-, and *Nitrobacter*, which converts NO_2^- to NO_3^-. From the global viewpoint, nitrogen that enters the soil ecosystem through natural processes rather than through artificial fertilization is derived from biological N_2 fixation and atmospheric deposition of NH_3, NH_4^+, and NO_3^-. Under conditions with no anthropogenic N deposition, more than 90% of plant nitrogen uptake is provided by internal cycling (Gosz 1981; Melillo 1981). The major pathways of N loss from soil ecosystems include bacterial denitrification, chemodenitrification, NH_3 volatilization, and leaching and erosion.

Although 79.1% of the atmosphere is composed of molecular nitrogen, much of it is unavailable to plants because of the difficulty associated with reduction or oxidation of the triple bond between the two N atoms. Those plants that develop symbiotic relationships with N-fixing procaryotes capable of producing nitrogenase to reduce N_2 to ammonia (bacteria *Rhizobia* and actinomycete *Frankia*) can benefit from the atmospheric N pool. A necessary precursor to symbiotic N_2 fixation in leguminous plants is the formation of nodules. Nodulation, however, is affected by soil acidity, and different leguminous plants have specific pH optima. Alexander (1980a) reported that symbiotic N_2 fixation does not occur below pH 4.6 for soybeans, 6.5 for alfalfa, and 5.2 for clover. Inhibition of N_2-fixing bacteria in soybean fields, using treatments with simulated rain of pH 3.0, was demonstrated by Miller et al. (1991). *Azotobacter* was usually confined to soils with pH >6.5, whereas some free-living bacteria (*Bacillus polymyxa* and *Beijerinckia* spp.) and some blue-green algae (*Anabaena* and *Nostoc* spp.) fix nitrogen at pH as low as 4.0 (Granhall and Selander 1973; Mulder and Brotonegoro 1974). Certain rhizobia, such as *Rhizobium meliloti*, were absent or deficient in soils below pH 5.5 (Bond 1974; Vincent 1974).

In nonagricultural systems, simulated acid rain of pH 3.2 inhibited *Rhizobium* nodulation (Shriner 1977), and N_2 fixation decreased in epiphytic lichens exposed to simulated acid rain of pH 4.0 or lower (Denison et al. 1977). Complete inhibition of N_2 fixation occurred at soil pH of 3.6 and 4.7, compared with no effects on samples incubated at pH 5.6 (Francis et al. 1980). Chang and Alexander (1983) demonstrated that the rate of nitrogen fixation in forest soils treated with simulated rain at pH 3.5 was significantly slower than in soils treated with rain at pH 5.6, and the inhibitory effect increased with amounts of rain at pH 3.5. In addition, Nohrstedt (1988) reported that liming of forest soil in central Sweden resulted in an increase in nitrogen fixation. However, Bitton and Boylan (1985) found no decrease in nitrogen fixation in three soil types after 92 and 690 d of exposure to acid rain of pH 3.7 and 3.0, compared with rain of pH 4.6. These results agree with those of Denison et al. (1977), who found that acidic deposition (pH of 5.0, 4.0, and 3.0) had little effect on N_2 fixation. These effects of acid precipitation on soil nitrogen cycling pathways do not appear to be of major significance in acidic soils (Francis 1982).

The decreased nitrogen-fixing activity of leguminous plants in response to acidity may also result from changes in the availability of nutrients or toxins as a consequence of soil acidification. Molybdenum, which is a necessary micronutrient for legumes, can be transformed into an inaccessible form at lower pH levels, whereas the potentially toxic elements Mn, Al, and Fe become more soluble at lower pH (Alexander 1980a).

Mineralization of organic forms of soil nitrogen to ammonia is mediated by various heterotrophic and autotrophic species of soil microorganisms under a wide range of soil pH, ranging from strongly alkaline to strongly acidic (Olson 1983). Because of the high biodiversity of ammonifying organisms, this step of nitrogen mineralization pathways is affected least by the different components of acidic deposition (Alexander 1977, 1980a,b). Dancer et al. (1973) reported that soil pH in the range 4.7–6.6 did not affect rates of ammonification appreciably, and Klein et al. (1984) demonstrated that ammonification in spodsols was not affected, even at pH 3.5. More-over, nitrogen mineralization in a deciduous forest soil was unaffected by irrigation with sulfuric and nitric acids for as long as 1 yr (Johnson and Todd 1984).

In contrast, nitrification is much more sensitive to acidic deposition, probably because the oxidation of NH_4^+ to NO_3^- is mediated by a restricted group of autotrophic microorganisms that have optimal growth and activity at pH 7.0–8.0 and are restricted at pH 5.7 (Alexander 1980b; Carlyle 1986; von Mersi et al. 1992; Tamm 1976). The activity of these microorganisms decreases rapidly as soil pH falls (Bitton et al. 1985; Brown 1985; De Boer et al. 1992; Francis 1982; Greszta et al. 1992; Johnson and Siccama 1983; McColl and Firestone 1987; Miller et al. 1991; Strayer et al. 1981; Stroo and Alexander 1986a,b). A positive correlation between rates of nitrifica-tion and soil pH in forest soils was also reported by Mladenoff (1987). However, Stevens and Wannop (1987) demonstrated that effective nitrifica-tion can take place in soil at pH below 4.0, and Johnson and Todd (1984) reported that nitrification in a deciduous forest soil was unaffected by irrigation with sulfuric and nitric acids for up to 1 yr. Similarly, no effects of simulated acid rain on nitrification were reported by McColl and Fire-stone (1991). This may indicate that formation of nitrate nitrogen (NO_3-N) in acidic soils is based on the activity of heterotrophic nitrifiers. Several authors have reported that heterotrophic nitrifiers are responsible for the formation of NO_3-N in strongly acidic soils (Adams 1986; von Mersi et al. 1992; Williams 1983). De Boer et al. (1989) observed acid-tolerant chemoli-thotrophic nitrification, which they attributed to the existence of previously unknown acid-tolerant, chemolithotrophic, ammonium-oxidizing bacteria. Stams et al. (1991) also indicated that nitrate formation from ammonium can occur even in an extremely acidic forest soil (pH < 4). An input of nitrogen together with acidic deposition was reported to have a negative effect on the numbers of nitrifiers and on nitrate mobilization (Verhoef et al. 1989).

Denitrification is catalyzed by a relatively diverse group of heterotrophic microorganisms and therefore is not usually very sensitive to acid precipita-tion (Firestone et al. 1984). Although denitrification was reported to be inhibited in extremely acidic soils, overall reaction rates have not been found to be affected by slight changes in soil pH (Firestone 1982), and some laboratory experiments have shown that the denitrification potential (a measure of denitrifier populations) increases after addition of simulated acid rain containing NO_3^- (McColl and Firestone 1987). Moreover, a change in soil pH can alter the composition of the products formed, thereby causing increased N_2O production relative to N_2 (Firestone et al. 1980). In soil slurries experiments (Struwe and Kjøller 1994), N_2O production was relatively uniform at different soil pH values, but total denitrification increased at higher soil pH. Increases in N_2O levels are not desirable because they may alter the atmospheric chemistry and have the potential to contribute to global climatic change (Crutzen 1970; Wang et al. 1976). Schmidt et al. (1988) estimated that nitrous oxide emission from temperate forest soils is the largest individual source of atmospheric N_2O.

Conflicting results have been reported from studies of the effects of acidic deposition on net nitrogen mineralization. Experimental acidification can stimulate nitrogen mineralization in soils (Persson et al. 1989; Strayer et al. 1981; Tamm 1976; Tamm et al. 1977), decrease it (Francis 1982; Klein et al. 1984) or have no effect (Johnson and Todd 1984; Moore 1987; Nohrstedt 1985). Liming of acid soils reportedly increases net nitrogen mineralization (Haynes and Swift 1988).

The mechanisms of the changes in nitrogen availability in response to acidity are not fully understood, but several authors (Agarwal et al. 1971; Heilman 1975; Johnson and Guenzi 1963) have suggested that they may be similar to those involved in the well-known "priming effect" or "salt" effect whereby additions of either fertilizer nitrogen or neutral salts can stimulate nitrogen mineralization. However, Martikainen (1985) attributed the inhibition of nitrification following salt (ammonium sulfate and potassium sulfate) treatments to a decrease in soil pH rather than to osmotic effects.

Tamm et al. (1977) hypothesized that an initial increase in net nitrogen mineralization after application of highly acidic treatments, in field experiments, could result from a partial mortality of soil microorganisms. This would cause a consequent release of biomass N, because the soil microflora biomass has a C:N ratio of about 5 (McGill et al. 1981) and can contain up to 4%–5% of the total nitrogen in forest soils (Persson 1983). However, Aber et al. (1982) considered that this was only a short-term phenomenon, depending on an unrealistically high dose of acidic treatments, and suggested that no long-term increases in N mineralization were likely to occur. They believed that a decrease in carbon mineralization following acidification of soil ecosystems would probably be accompanied by corresponding decreases in net nitrogen mineralization. This hypothesis was supported by Klein et al. (1984), who reported a decrease in both C and N mineralization in response to soil acidification.

Laboratory experiments by Persson et al. (1989) demonstrated that strong soil acidification increased net N mineralization significantly. However, the mineralization of biomass N in dead soil organisms was sufficient to explain only 25%–30% of the increase in net N mineralization. Most of the effects of acidification on N mineralization resulting from the acid and lime treatments depended on the fact that acidification reduced and liming increased the availability of C and N to microorganisms. Acidification reduced the incorporation of N compounds from dead organisms into soil organic matter, resulting in an increased availability of substances rich in N to the microorganisms, and thus increased net soil N mineralization.

Overall, most natural soil ecosystems are efficient at retaining N through internal cycling and N losses as a result of leaching and denitrification are very small. In recent years, however, increased leaching of N in the form of nitrate has been observed from forest soils in areas with high N deposition (Nilsson and Grennfelt 1988; Hauhs et al. 1989). This indicates that such forest ecosystems may become "N-saturated" from atmospheric N deposition. Gundersen (1991) estimated that a range of 2–20 kg N ha^{-1} yr^{-1} represents a critical N load for many forest ecosystems. Thus, input of nitrogen with acidic deposition and from other sources, in excess of plant demand, may cause nutritional imbalances and leaching of nutrients, promoting further soil acidification.

The Sulfur Cycle.   Sulfur is an essential macronutrient required by plants and animals in virtually all ecosystems, although it is needed in lesser amounts than nitrogen. The overall biological demand for S is less than 5% of that for N, on a molar basis, and this is usually exceeded by atmospheric S deposition in many areas where this occurs (Johnson 1984). This probably explains the apparent lack of interest and paucity of studies on the effects of acidic deposition on sulfur cycling.

The cycling of sulfur (Binkley et al. 1989; Germida et al. 1992; Johnson et al. 1986; Stevenson 1986) resembles the cycling of nitrogen in many ways. It differs in that elemental S is metabolized easily by soil microorganisms and is a relatively unimportant source of sulfur in the functioning of most ecosystems. Unlike nitrate, sulfate can adsorb onto soils that are rich in Fe and Al oxides and poor in organic matter (Johnson 1987).

Studies about the sulfur inputs from acidic deposition to soil ecosystems have shown detrimental effects of S on some dynamic soil processes (Alexander 1980a; Babich and Stotzky 1978; Bewley and Parkinson 1984; Bewley and Stotzky 1983a,b; Bryant et al. 1979; Germida et al. 1992; Tabatabai 1985). However, there is no agreement on the extent of negative impact of this form of pollution on different ecosystems. Transformations of sulfur in soil organic matter, to forms available to plants, is a strictly microbiological process (Stevenson 1986), and hence the turnover of sulfur through mineralization–immobilization processes is affected by the growth and activity of soil microorganisms. Consequently, changes in the soil chemical environment that are associated with acidic deposition, such as changes in soil pH, mobilization of toxic metals, and the availability of nutrients that can influence the growth and activity of soil microorganisms, can also affect sulfur cycling in soils.

Patterns of sulfur immobilization–mineralization dynamics in decomposing leaf litter can also be affected by ambient sulfate concentrations. A higher SO_4^{2-} content in artificial acid rain resulted in greater retention of S in decomposing birch leaf litter (Hågvar and Kjøndal 1981a) and spruce litter (Meiwes and Khanna 1981) and in greater immobilization of S in microbial biomass (Hern et al. 1985), and in higher C-bonded S concentrations in those soil microorganisms involved in leaf litter decomposition (Saggar et al. 1981). The addition of sulfur also resulted in a highly significant reduction in S mineralization in two different soils (Ghani et al. 1992). However, no increases in S content of leaf litter were reported for several hardwood species exposed to elevated sulfur levels (Lee and Weber 1983), and no effects of moderately high sulfate additions (4 mg S L^{-1}) on S retention in Douglas-fir and red alder leaf litters were reported by Homann and Cole (1990). These authors concluded that ambient SO_4^{2-} concentrations rarely influence the amount of insoluble organic S formed and retained in decomposing litter.

Soils that are affected by large amounts of acidic deposition invariably have a higher total S content, of which more than 95% is in organic form. Mineralization of these sulfur compounds may be a more important source of SO_4^{2-} than that from exogenous inputs. Wainwright and Nevell (1984) reported an increase in inorganic sulfur, in forms such as SO_4^{2-}, SO_3^{2-}, and $S_2O_3^{2-}$, in soils receiving large doses of acidic deposition. Such increases in inorganic forms of S are usually associated with increases in the numbers and species of S-oxidizing microorganisms (Germida et al. 1992; Wainwright 1979).

The relative size of the sulfate pool is considered to play an important role in controlling the rates of organic S mobilization in any given soil (Stanko and Fitzgerald 1990). Lettl (1981a,b) reported that thiobacilli are usually absent or occur only at low frequencies in non-sulfur-polluted soils, whereas their numbers increase substantially in soils exposed to sulfur deposition. Miller et al. (1991) reported increases in thiosulfate-oxidizer microbial populations in soils treated with simulated acid rain (see Table 3). They attributed this to an increase in amounts of thiosulfate, resulting from an assimilatory sulfate reduction to cysteine and methionine followed by mineralization to sulfide. The sulfide released could then be used by S-oxidizing species. Additional evidence for the effects of acidic deposition on sulfur transformations in soil ecosystems may be the increase in rhodanase activity in sulfur-polluted soils (Wainwright 1979). Arylsulfatase activity in peat has been shown to correlate directly with increasing levels of atmospheric S deposition (Jarvis et al. 1987). However, microbiological examination of thin sections of S-treated soil indicated that the percentage of oxidized S decreased significantly after increasing the rate of S application (Modaihsh et al. 1989; Rida and Modaihsh 1988).

Little quantitative information is available on sulfur budgets in ecosystems situated in areas with increased amounts of acidic deposition. Johnson et al. (1982) reported that the sulfur requirement of a chestnut oak forest in the southern U.S. was about 22 kg S ha^{-1} annually, with approximately 10% of this supplied by recycling from senescent leaves, and an annual accumulation of S in biomass of about 2.1 kg S ha^{-1}. Given a rate of acidic deposition at time of sampling of about 5 kg S ha^{-1} annually, it was apparent that S deposition was exceeding the rate of S accumulation in microbial biomass for most forests in the region. Several investigators have also reported the accumulation of S in different forest ecosystems (Cleaves et al. 1974; Johnson et al. 1985; Meiwes and Khanna 1981; Shriner and Henderson 1978; Swank and Douglass 1977; Waide and Swank 1987; Weller et al. 1986).

In other studies, sulfur leaching outputs were equal to or exceeded inputs (Cole and Johnson 1977; Likens et al. 1977). Johnson et al. (1980) proposed that these differences in S accumulation in ecosystems were the result of abiotic sulfate adsorption to Fe and Al oxides in subsurface soil horizons, although biological retention could also be an important mechanism of removal of S from both organic and mineral soil horizons (vanLoon et al. 1987). Johnson et al. (1982) suggested that increased deposition of S onto soil ecosystems from air pollution could cause a change in the nature of forest sulfur cycling from a biogeochemical cycle to a predominantly geochemical cycle. A summary of the current literature on budgets of S inputs and outputs for several intensively studied sites in the U.S and Canada was compiled by Rochelle et al. (1987). They concluded that S retention could be related to both soil type and even the extent of the last glaciation, and that regional variations in sulfur retention in North America are important in determining and predicting the overall effects of acidic deposition in those regions.

In general, the net effect of acidic deposition on sulfur cycling with regard to ecosystem productivity depends on a number of site-specific factors, particularly nutrient status and amount of acidic input. In S- and N-limited ecosystems with high contents of base cations, moderate inputs of acidic deposition can increase plant growth, whereas high levels of acidic deposition may decrease productivity in ecosystems with sufficient N and S but inadequate amounts of cation nutrients.

The Phosphorus Cycle.   The soil phosphorus cycle resembles the nitrogen cycle. After nitrogen, phosphorus is the mineral nutrient required most by plants and microorganisms. Phosphorus can occur in the rhizosphere in a variety of organic and inorganic forms, which are released slowly into the soil solution. Organic forms of P account for 20%-90% of the total P in the surface horizon of most soils. In natural undisturbed ecosystems, the P cycle is tightly closed, and most of the P taken up by plants is supplied from recycling of plant residue P by soil microorganisms (Tate 1984). Because much of the turnover of organic P is controlled by the activities of soil microorganisms, any changes in the soil microbial community that result from the effects of various components of acidic deposition will also affect the P mineralization–immobilization dynamics in the relevant ecosystem.

Very little experimental information is available on the effects of acidic deposition on P cycling (Cole and Stewart 1983). It is generally accepted that decreases in soil pH reduce availability of inorganic phosphorus by causing precipitation of aluminum phosphate or iron phosphates, thereby increasing the importance of recycling organic P (Mohren et al. 1986; Myr-old 1987). Decreases in plant P uptake in acidified soils (pH 3.0 and 4.0) were reported by McColl and Firestone (1991). However, no effects of acidic deposition on the amount of extractable P or phosphorus solubility were reported.

Changes in the cycling of soil P may result from any nutrient imbalances brought about by considerable increases of minerals in atmospheric deposition. Mohren et al. (1986) suggested that in nitrogen-poor sites the elimination of nitrogen deficits caused by acidic deposition can increase plant growth and tissue N content up to limits imposed by the availability of some other element, such as phosphorus. The increased demand for P may then lead to a decrease in concentrations of the available forms of P and to an overall P deficiency in the ecosystem. Positive reactions to phosphorus fertilization resulting from this deficiency have already been reported from the Netherlands (Van den Burg 1976).

In summary, the mineralization of nutrients contained in soil organic matter to forms available for plant uptake is a microbiological process that is controlled by the rates of growth and activity of soil microorganisms. Consequently, changes in the soil chemical environment that are associated with acidic deposition, such as changes in soil pH, mobilization of toxic metals, and the availability of nutrients that can influence growth and activity of soil microorganisms, can affect nutrient cycling in soils.

Summary

Research into the environmental impact of acidic deposition, which began in 1852, has become much more intensive in recent years. For instance, there is now an extensive European Air Chemistry Network and a European Monitoring and Evaluation Program to evaluate the extent of the problem there. Acidic deposition in Europe increased by about 50% from the late 1950s to the late 1960s, then declined in the 1970s and 1980s. Monitoring of acidic deposition began in North America in the 1970s and, beginning in 1979, a 10-yr monitoring program was established. There have been significant reductions in sulfur and nitrogen emissions from industrial sources and consequent acid precipitation in the U.S. during the past two decades.

Acidic deposition relates to the total amounts of acids and acidifying compounds present in both wet and dry atmospheric deposition. Wet acidic deposition consists mainly of sulfuric and nitric acids. Dry deposition consists mainly of SO_2, which can be oxidized further to sulfuric acid, NO and NO_2, and volatile organic compounds. The areas with maximum acidic deposition in North America are the northeastern U.S. and southeastern Canada.

Most environmental attention has been on the effects of acidic deposition on forest trees and aquatic ecosystems, with relatively little attention being paid to soil ecosystems. The aim of this review has been to survey comprehensively all available data on the effects of acidic deposition on soil organisms and processes in terrestrial ecosystems.

It is feasible that the overall effects of acidic deposition on soil ecosystems could be beneficial as well as detrimental, depending on the chemical and physical properties of the soil. However, it is much more likely that acidic deposition will have adverse effects by decreasing the pH of soils to levels below the optimum for growth and reproduction of plants and functioning of the soil decomposer community. Many soils however have considerable buffering capacity that could help to minimize the overall effects of acidic deposition.

Soil-inhabiting invertebrates are essential contributors to the decomposer food web and nutrient cycling pathways. Most studies that have assessed the effects of acidic deposition on soil invertebrate populations and communities have focused on the most abundant groups—Collembola, Acari, and Enchytraeidae. Most experimental studies have involved application of sulfuric acid solutions, sometimes in combination with nitric acid, to soils to assess their effects. Other studies have investigated the effects of liming, nitrogenous fertilizers, acidic and neutral sulfate solutions, and fumigation with SO_2 on soil-inhabiting invertebrates. Most experiments reported in the literature were short term. Most of the available data suggest that induced acidification of soils leads to changes in soil invertebrate communities in favor of species that are dominant in naturally acidic soils. However, the data are extremely variable. Lower pH levels tend to decrease numbers of Protozoa, Rotifera, and Enchytraeidae. By contrast, populations of many species of Collembola and oribatid mites tended to increase after acidification. Low doses of acid favored earthworm populations, but larger doses decreased them. Populations of many other soil-inhabiting invertebrates decreased in response to acidification. The effects of acidic deposition on invertebrate populations may be direct, by changing the soil environment, or indirect, through its influence on populations of microorganisms upon which the invertebrates feed.

The effects of acidic deposition on soil microorganisms are variable and are probably greatest on soil bacteria, which have an optimum pH near neutral; fungi, which have optimal pH levels about 1–2 pH units lower, are influenced less. The effects of acidification on microorganisms may be through increased concentrations of anionic components of strong acids, mobilization of toxic metals, or changes in interactions with other microorganisms or invertebrates. There is considerable evidence that acidic deposition can have adverse effects on overall soil microbial biomass. Although soil fungi tend to be more tolerant of acidic deposition than bacteria, a number of workers reported decreased populations of saprophytic fungi in response to acidification. Similarly, sulfur dioxide has been reported to have adverse effects on soil fungi. Some mycorrhizal fungi seem to be susceptible to acidic deposition, but others are not. Most workers studying ectomycorrhizal fungi reported that acidic deposition adversely affected them. There is some evidence that larger doses of some forms of nitrogen have adverse effects on mycorrhizal fungi.

There have been relatively few good studies on the effects of acidic deposition on dynamic soil processes; about half of these reported no effects of acidification on organic matter breakdown. Similarly, studies into the effects of acidic deposition on soil respiration yielded inconsistent results, although decrases in CO_2 production after acidification were reported in a number of studies.

Although there has been considerable research on soil enzymes, there have been relatively few studies of the effects of acidification. Different soil enzymes differ greatly in their response to acidic deposition, but many of them seem to be sensitive to acidification, particularly at high doses.

Nitrification appears to be much more sensitive to acidic precipitation than some of the other soil processes, and many workers reported progressive inhibition of this process as a result of acidification. By contrast, ammonification, which is catalyzed by a relatively diverse group of heterotrophic organisms, is not nearly as sensitive to acid precipitation. There is some evidence of an initial stimulation of net N mineralization in response to acid precipitation, followed by a later depression.

Acid precipitation, particularly that containing SO_4^{2-}, can influence S mineralization in the sulfur cycle. Soils affected by large quantities of acidic deposition almost invariably have a higher total S content, of which more than 95% is in the organic form. It has been suggested that increased deposition of S on soil ecosystems can cause soil changes from a biogeochemical cycle to a predominantly geochemical cycle. Sulfur retention may be related both to soil order and to the extent of glaciation.

There is very little information on the effects of acidic deposition on phosphorus cycling. However, most evidence concludes that the resulting decreases in pH reduce the availability of inorganic P and increase the importance of recycling organic forms of P.

Overall, reported changes in soil ecosystems affected by acid deposition appear to be a multiple-factor response syndrome. The outcome of these changes was characterized as "cosystem destabilization." This destabilization may lead to a new equilibrium that will integrate new rates of retention and mineralization of nutrients, resulting in an alteration of ecosystem productivity.

References

Aber JD, Hendrey GR, Botkin DB, Francis AJ, Melillo JM (1982) Water Air Soil Pollut 18:405.

Abrahamsen G (1970) Forest fertilization and the soil fauna. Tidsskr Skogbruk 78: 296–303.

Abrahamsen G (1972a) Ecological study of Enchytraeidae (Oligochaeta) in Norwegian coniferous forest soils. Pedobiologia 12:26–82.

Abrahamsen G (1972b) Ecological study of Lumbricidae (Oligochaeta) in Norwegian coniferous forest soils. Pedobiologia 12:267–281.

Abrahamsen G, Bjor K, Teigen O (1976) Field experiments with simulated acid rain in forest ecosystems. FR 4/76. SNSF, Olso-As, Norway.

Abrahamsen G, Hovland J, Hågvar S (1980) Effects of artificial acid rain and liming on soil organisms and the decomposition of organic matter. In: Hutchinson TC, Havas M (ed) Effects of Acid Precipitation on Terrestrial Ecosystems. Plenum Press, New York, pp 341–362.

Abrahamsen G (1983) Effects of lime and artificial acid rain on the enchytraeid (Oligochaeta) fauna in coniferous forest. Holarct Ecol 47:787–803.

Adams JA (1986) Identification of heterotrophic nitrification in strongly acidic larch humus. Soil Biol Biochem 18:339–341.

Adams MB, O'Neill EG (1991) Effects of ozone and acidic deposition on carbon allocation and mycorrhizal colonization of *Pinus taeda* L. seedlings. For Sci 37: 5–16.

Aerts R, van Logtestijn R, van Staalduinen M, Toet S (1995) Nitrogen supply effects on productivity and potential leaf litter decay of *Carex* species from peatlands differing in nutrient limitation. Oecologia 104:447–453.

Agarwal AS, Singh BR, Kanehiro Y (1971) Ionic effects of salt on mineral release in an allophanic soil. Soil Sci Soc Am Proc 35:454–457.

Aguilar G, Trejo BA, García JM, Huitrón C (1991) Influence of pH on endo- and exo-pectinase production by *Aspergillus* sp. CH-Y-1043. Can J Microbiol 37: 912–917.

Alexander M (1977) Introduction to Soil Microbiology. Wiley, London.

Alexander M (1980a) Effects of acidity on microorganisms and microbial processes in soil. In: Hutchinson TC, Havas M (eds) Effects of Acid Precipitation on Terrestrial Ecosystems. Plenum Press, New York, pp 363–374.

Alexander M (1980b) Effects of acid precipitation on biochemical activities in soil. In: Drabloes D, Tollan A (eds) Ecological Impact of Acid Precipitation. Proceedings of an International Conference. SNSF, Oslo, Norway.

Ammer S, Makeschin F (1994) Auswirkungen experimenteller saurer Beregnung und Kalkung auf die Regenwurmfauna (*Lunbricidae, Oligochaeta*) und die Humusform in einem Fichtenaltbestand (Höglwaldexperiment). Forstwiss Centralbl (Hamb) 13:70–85.

Andersson FT, Fagerstrom T, Nilsson SI (1980) Forest ecosystem responses to acid deposition–hydrogen ion budget and nitrogen/tree growth model approaches. In: Hutchinson TC, Havas M (eds) Effects of Acid Precipitation on Terrestrial Ecosystems. Plenum Press, New York, pp 319–334.

Andersson S, Söderström B (1995) Effects of lime ($CaCO_3$) on ectomycorrhizal colonization of *Picea abies* (L.) Karst. seedlings planted in a spruce forest. Scand J For Res 10:149–154.

Anonymous (1988) Luchtkwaliteit, jaarverslag 1987. Rapp. 228703002. RIVM, Bithoven.

Antibus RK, Linkins AE III (1992) Effects of liming a red pine forest on mycorrhizal numbers and mycorrhizal and soil acid activities. Soil Biol Biochem 24:479–487.

Armentano T, Loucks O (1990) Spatial patterns of S and N deposition in the midwestern hardwoods region. In: Loucks OL (ed) Air Pollutants and Forest Response: The Ohio Corridor Study. Year-4 Annual Report. Miami University, Holcomb Research Institute, Oxford, OH, pp 54–84.

Arnebrant K, Bääth E, Soederstroem B (1990) Changes in microfungal community structure after fertilization of Scots pine forest soil with ammonium nitrate or urea. Soil Biol Biochem 22:309–312.

Artemjeva TI, Gatilova FG (1975) Soil microfauna changes under the influence of various fertilizers. In: Vanek J (ed) Progress in Soil Zoology. Academia, Prague, pp 463–468.

Ashraf M (1969) Studies on the biology of Collembola. Rev Ecol Biol Soil 6:337–347.

Axelsson B, Lohm U, Lundkvist H, Persson T, Skoglund J, Wiren A (1973) Effects of nitrogen fertilization on the abundance of soil fauna populations in a Scots pine stand. Research Notes 14. Institute Växtekologi, Marklära.

Bääth E, Lohm U, Lundgren B, Rosswall T, Soderstrom B, Sohlenius B, Wiren A (1978) The effect of nitrogen and carbon supply on the development of soil organism populations and pine seedlings: a microcosm experiment. Oikos 31: 153–163.

Bääth E, Lundgren B, Soderstrom B (1979) Effects of artificial acid rain on microbial activity and biomass. Bull Environ Contam Toxicol 23:737–740.

Bääth E, Berg B, Lohm U, Lundgren B, Lundkvist H, Rosswall T, Soderstrom B, Wiren A (1980) Effects of experimental acidification and liming on soil organisms and decomposition in a Scots pine forest. Pedobiologia 20:85–100.

Bääth E, Lundgren B, Soderstrom B (1984) Fungal populations in podzolic soils experimentally acidified to simulate acid rain. Microb Ecol 10:197–203.

Bääth E, Arnebrant K (1994) Growth rate and response of bacterial communities to pH in limed and ash-treated forest soils. Soil Biol Biochem 26:995–1001.

Babich H, Stotzky G (1978) Atmospheric sulfur compounds and microbes. Environ Res 15:513–531.

Babich H, Stotzky G (1980) Environmental factors that influence the toxicity of heavy metals and gaseous pollutants to microorganisms. Crit Rev Microbiol 8: 99–145.

Badalucco L, Greco S, Dell'Orco S, Nannipieri P (1992) Effects of liming on some chemical, biochemical, and microbiological properties of acid soils under spruce (*Picea abies* L.). Biol Fertil Soils 14:76–83.

Baker EW, Wharton GW (1952) An Introduction to Acarology. Macmillan, New York.

Bassus W (1960) Die Nematodenfauna des Fichtenrohhumus unter dem Einflub der Kalkdungung. Nematologica 5:86–91.

Bassus W (1967) Der Einflub von Meliorations und Dungungsmassnahmen auf die Nematodenfauna verschiedener Waldenboden. Pedobiologia 7:280–295.

Behan VM, Hill S, Kevan DKM (1978) Effects of nitrogen fertilizers, as urea, on *Acarina* and other arthropods in Quebec black spruce humus. Pedobiologia 18: 249–263.

Berg B (1986a) The influence of experimental acidification on needle litter decomposition in a *Picea abies* L. forest. Scand J For Res 1:317–322.

Berg B (1986b) The influence of experimental acidification on nutrient release and decomposition rates of needle and root litter in the forest floor. For Ecol Manage 15:195-213.

Berger H, Foissner W, Adam H (1986) Field experiments on the effects of fertilizers and lime on the soil microfauna of an alpine pasture. Pedobiologia 29:261-272.

Bewley RJ, Stotzky G (1983a) Simulated acid rain (H_2SO_4) and microbial activity in soil. Soil Biol Biochem 15:425-429.

Bewley RJ, Stotzky G (1983b) Anionic constituents of acid rain and microbial activity in soil. Soil Biol Biochem 15:431-437.

Bewley RJ, Parkinson D (1984) Effects of sulphur dioxide pollution on forest soil microorganisms. Can J Microbiol 30:179-185.

Bewley RJ, Stotzky G (1984) Degradation of vanillin in soil-clay mixtures treated with simulated acid rain. Soil Sci 137:415-418.

Bewley RJ, Parkinson D (1985) Bacterial and fungal activity in sulfur dioxide polluted soils. Can J Microbiol 31:13-15.

Bewley RJ, Parkinson D (1986a) Sensitivity of certain soil microbial processes to acid deposition. Pedobiologia 29:73-84.

Bewley RJ, Parkinson D (1986b) Monitoring the impact of acid deposition on the soil microbiota, using glucose and vanillin decomposition. Water Air Soil Pollut 27:57-68.

Beyer WN, Hensler G, Moore J (1987) Relation of pH and other soil variables to concentration of Pb, Cu, Zn, Cd, and Se in earthworms. Pedobiologia 30:167-172.

Bigg WL (1981) Some effects of nitrate, ammonium and mycorrhizal fungi on the growth of Douglas fir and Sitka spruce. Ph.D. thesis, University of Aberdeen, Scotland.

Binkley D, Richter D (1987) Nutrient cycles and H^+ budgets of forest ecosystems. In: Macfayden A, Ford ED (eds) Advances in Ecological Research, Vol. 16. Academic Press, London, pp 1-51.

Binkley D, Driscoll CT, Allen HL, Schoeneberger P, McAvoy D (1989) Acidic Deposition and Forest Soils. Ecological Studies, Vol. 72. Springer-Verlag, New York.

Bitton G, Boylan RA (1985) Effects of acid precipitation on soil microbial activity: I. Soil core studies. J Environ Qual 14:66-68.

Bitton G, Volk BG, Graetz DA, Bossart JM, Boylan RA, Byers GE (1985) Effects of acid precipitation on soil microbial activity: II. Field studies. J Environ Qual 14:69-71.

Björkman E (1942) Über die Bedingungen der Mykorrhizabildung bei Kiefer und Fichte. Symb Bot Ups 6:1-191.

Blackmer AM, Bremner JM, Schmidt EL (1980) Production of nitrous oxide by ammonia-oxidizing chemoautotrophic microorganisms in soil. Appl Environ Microbiol 40:1060-1066.

Blair JM, Parmelee RW, Wyman RL (1994) A comparison of the forest floor invertebrate communities of four forest types in the northeastern U.S. Pedobiologia 38:146-160.

Blaschke H (1988) Mycorrhizal infection and changes in fine-root development of Norway spruce influenced by acid rain in the field. In: Ectomycorrhiza and Acid Rain. Proceedings, Workshop Ectomycorrhiza/Expert Meeting, Berg en Dal, Netherlands. CEC Air Pollut Res Rep 12 (EUR 11534):112-115.

Blaschke H (1990) Mycorrhizal populations and fine root development on Norway spruce exposed to controlled doses of gaseous pollutants and simulated acidic rain treatments. Environ Pollut 68:409–417.

Bohlen PJ, Edwards CA (1994) The response of nematode trophic groups to organic and inorganic nutrient inputs in agroecosystems. Defining Soil Quality for a Sustainable Environment, Soil Sci Soc Amer Spec Publ No 35 pp 235–244.

Bond G (1974) Symbiosis with actinomycete-like organisms. In: Quispel A (ed) The Biology of Nitrogen Fixation. Plenum Press, Amsterdam, pp 342–378

Booth IR (1985) Regulation of cytoplasmic pH in bacteria. Microbiol Rev 49:359–378.

Borror DJ, DeLong DM, Triplehorn CA (1976) Introduction to the study of insects. Holt, Rinehart and Winston, New York.

Brewer PF, Heagle AS (1983) Interactions between *Glomus geosporum* and exposure of soybeans to ozone or simulated acid rain in the field. Phytopathology 73:1035–1040.

Brown K (1985) Acid deposition: Effects of sulphuric acid at pH 3 on chemical and biochemical properties of bracken litter. Soil Biol Biochem 17:31–38.

Bryant RD, Gordy EA, Laishley EJ (1979) Effect of soil acidification on the soil microflora. Water Air Soil Pollut 11:437–445.

Brzeski MW, Dowe A (1969) Effect of pH on *Tylenchorhynchus dubius* (Nematoda, Tylenchidae). Nematologica 15:403–407.

Burgess RL (ed) (1984) Effects of acidic deposition on forest ecosystems in the Northeastern United States: an evaluation of current evidence. Institute of Environmental Program Affairs, State University of New York, Syracuse.

Burns RG (1978) Soil Enzymes. Academic Press, London.

Burt AJ, Hashem AR, Shaw G, Read DJ (1986) 1st SEM, Dijon, 1-5.7.1985. INRA, Paris, pp 683–687.

Campbell CD, Chapman SJ, Urquhart F (1995) Effect of nitrogen fertilizer on temporal and spatial variation of mineral nitrogen and microbial biomass in a silvopastoral system. Biol Fertil Soils 19:177–184.

Carlyle JC (1986) Nitrogen cycling in forested ecosystems. For Abstr 47:307–336.

Carter MR (1986) Microbial biomass and mineralizable nitrogen in solonetzic soils: influence of gypsum and lime amendments. Soil Biol Biochem 18:531–537.

Chambers CA, Smith SE, Smith FA (1980) Effects of ammonium and nitrate ions on mycorrhizal infection, nodulation and growth of *Trifolium subterraneum*. New Phytol 85:47–62.

Chang F-H, Alexander M (1983) Effect of simulated acid precipitation on algal fixation of nitrogen and carbon dioxide in forest soils. Environ Sci Technol 17:11–13.

Chang F-H, Alexander M (1984) Effect of simulated acid precipitation on decomposition and leaching of organic carbon in forest soils. Soil Sci 138:226–234.

Chétail M, Krampitz G (1982) Calcium and skeletal structures in mollusc: concluding remarks. In: Proceedings of the 7th International Malacological Congress. Malacologia 22:337–339.

Clapperton MJ, Parkinson D (1990) Effects of SO_2 on VA Mycorrhizae associated with a sub-montane mixed grass prairie in Alberta, Canada. Can J Bot 68:1646–1650.

Clapperton MJ, Reid DM, Parkinson D (1990) Effects of sulfur dioxide fumigation on *Phieum pratense* and vesicular-arbuscular mycorrhizal fungi. New Phytol 115:465–469.

Clapperton MJ, Reid DM (1992) Effects of low-concentration sulfur dioxide fumigation and vesicular-arbuscular mycorrhizas on ^{14}C-partitioning in *Phleum pratense* L. New Phytol 120:381–387.

Cleaves ET, Fisher DW, Bricker OP (1974) Chemical weathering of serpentinite in the eastern Piedmont of Maryland. Geol Soc Am Bull 85:437–444.

Cole CV, Stewart JW (1983) Impact of acid deposition on P cycling. Environ Exp Bot 23:235–241.

Cole DW, Johnson DW (1977) Atmospheric sulfate additions and cation leaching in Douglas-fir ecosystem. Water Resour Res 13:313–317.

Cowling E (1982) An historical resume of progress in scientific and public understanding of acid precipitation and its biological consequences. In: D'itri FM (ed) Acid Precipitation: Effects on Ecological Systems. Ann Arbor Science, Ann Arbor, MI, pp 43–83.

Craft CB, Webb JW (1984) Effects of acidic and neutral sulfate salt solutions on forest floor arthropods. J Environ Qual 13:436–440.

Cronan CS (1985) Comparative effects of precipitation acidity on three forest soils: carbon cycling responses. Plant Soil 88:101–112.

Crowell HH (1973) Laboratory study of calcium requirements of the brown garden snail *Helix aspersa* Müller. Proc Malacol Soc London 40:491–503.

Crutzen PJ (1970) The influence of nitrogen oxides on the atmospheric ozone content. QJR Meteorol Soc 96:320–325.

Dähne J, Klingelhöfer D, Ott M, Rothe GM (1995) Liming induced stimulation of the amino acid metabolism in mycorrhizal roots of Norway spruce (*Picea abies* [L.] Karst.). Plant Soil 173:67–77.

Dancer WS, Peterson LA, Chesters G (1973) Ammonification and nitrification of nitrogen as influenced by soil pH and previous nitrogen treatments. Soil Sci Soc Am Proc 37:67–69.

Davidson EA, Myrold DD, Groffman PM (1990) Denitrification in a temperate forest ecosystem. In: Gessel SP, Lacate DS, Weetman GF, Powers RF (eds) Sustained Productivity of Forest Soils University of British Columbia, Vancouver, pp 196–220.

Dean RA, Timberlake WE (1989a) Production of cell wall-degrading enzymes by *Aspergillus nidulans*: a model system for fungal pathogenesis of plants. Plant Cell 1:265–273.

Dean RA, Timberlake WE (1989b) Regulation of the *Aspergillus nidulans* pectate lyase gene (*pelA*). Plant Cell 1:275–284.

De Boer W, Gunnewiek PJAK, Troelstra SR, Laanbroek HJ (1989) Two types of chemolithotrophic nitrification in acid heathland humus. Plant Soil 119:229–235.

De Boer W, Tietema A, Gunnewiek PJAK, Laanbroek HJ (1992) The chemolithotrophic ammonium-oxidizing community in a nitrogen-saturated acid forest soil in relation to pH-dependent nitrifying activity. Soil Biol Biochem 24:229–234.

de Goede RGM, Dekker HH (1993) Effects of liming and fertilization on nematode communities in coniferous forest soils. Pedobiologia 37:193–209.

Denison R, Caldwell B, Bormann B, Eldred L, Swanberg C, Anderson S (1977) The effects of acid rain on nitrogen fixation in western Washington coniferous forest. Water Air Soil Pollut 8:21–34.

Dighton J, Mason PA (1985) Mycorrhizal dynamics during forest tree development. In: Moore D, Casselton LA, Wood DA, Frankland JC (eds) Developmental Biology of Higher Fungi. Cambridge University Press, Cambridge, pp 117–139.

Dighton J, Skeffington RA (1987) Effects of artificial acid precipitation on the mycorrhizas of Scots pine seedlings. New Phytol 107:191–202.

Dighton J, Jansen J (1991) Atmospheric pollutants and ectomycorrhizae: more questions than answers? Environ Pollut 73:179–203.

Dignon J (1992) NO_x and SO_x emissions from fossil fuels: a global distribution. Atmos Environ 26A:1157–1163.

Dillon PJ, Lusis M, Reid R, Yap D (1988) Ten-year trend in sulphate, nitrate, and hydrogen deposition in central Ontario. Atmos Environ 22:901–905.

Dmowska E (1993) Effects of long-term artificial acid rain on species range and diversity of soil nematodes. Eur J Soil Biol 29:97–107.

Dmowska E (1995) Influence of simulated acid rain on communities of soil nematodes. Acta Zool Fenn 196:321–323.

Dodd JL, Lauenroth WK (1981) Effects of low-level SO_2 fumigation on decomposition of western wheatgrass litter in a mixed grass prairie. Water Air Soil Pollut 15:257–261.

Domanski S, Kowalski T (1987) Fungi occurring on forests injured by air pollutants in the Upper Silesia and Cracow industrial regions. x. Mycoflora of dying young trees of *Alnus incana*. Eur J For Pathol 17:337–348.

Driscoll CT, Likens GE (1982) Hydrogen ion budget of an aggrading forest watershed. Tellus 34:283–292.

Edwards CA, Lofty JR (1969) The influence of agricultural practice on soil microarthropod populations. In: Sheals GJ (ed) The Soil Ecosystem. Systematics Assciation, London, pp 237–247.

Edwards CA, Lofty JR (1975) The invertebrate fauna of the Park Grass Plots. I. Soil fauna. Rothamsted Exp Stn Rep 1974 (Part 2): 133–154.

Edwards CA, Butler CG, Lofty JR (1976) The invertebrate fauna of the Park Grass Plots. II. Surface fauna. Rothamsted Exp Stn Rep 1975 (Part 2): 133–154.

Edwards CA, Bohlen PJ (1996) The Biology and Ecology of Earthworms, 3rd Edition. Chapman and Hall, London.

Edwards GS, Kelly JM (1992) Ectomycorrhizal colonization of loblolly pine seedlings during three growing seasons in response to ozone, acidic precipitation, and soil Mg status. Environ Pollut 76:71–77.

Entry JA, Cromack K Jr, Stafford SG, Castellano MA (1987) The effect of pH and aluminum concentration on ectomycorrhizal formation in *Abies balsamea*. Can J For Res 17:865–871.

Entry JA, Backman CB (1995) Influence of carbon and nitrogen on cellulose and lignin degradation in forest soils. Can J For Res 25:1231–1236.

Erland S, Söderström B (1991) Effects of liming on ectomycorrhizal fungi infecting *Pinus sylvestris* L. III. Saprophytic growth and host plant infection at different pH values by some ectomycorrhizal fungi in unsterile humus. New Phytol 117: 405–411.

Escritt JR, Arthur JH (1948) Earthworm control—a résumé of methods available. J Board of Greenkeep Res 7:49.

Escritt JR, Lidgate HR (1964) Report on fertilizer trials. J Sports Turf Res Inst 40: 7–42.

Esher RJ, Marx DH, Ursic SJ, Baker RL, Brown LR, Coleman DC (1992) Simulated acid rain effects on fine roots, ectomycorrhizae, microorganisms, and invertebrates in pine forests of the southern United States. Water Air Soil Pollut 61:269–278.

Faber JH (1991) Functional classification of soil fauna: a new approach. Oikos 62: 110–117.

Faber JH, Verhoef HA (1991) Functional differences between closely-related soil arthropods with respect to decomposition processes in the presence or absence of pine tree roots. Soil Biol Biochem 23:15–23.

Falappi D, Farini A, Ranalli G, Sorlini C (1994) Effects of simulated acid rain on some microbiological parameters of subacid soil. Chemosphere 28:1087–1095.

Fauci MF, Dick RP (1994) Soil microbial dynamics: short- and long-term effects of inorganic and organic nitrogen. Soil Sci Soc Am J 58:801–806.

Fellows PJ, Worgan JT (1984) An investigation into the pectolytic activity of the yeast *Saccharomycopsis fibuliger*. Enzyme Microb Technol 6:405–410.

Fenn ME, Dunn PH, Durall DM (1989) Effects of ozone and sulfur dioxide on phyllosphere fungi from three tree species. Appl Environ Microbiol 55:412–418.

Firestone MK, Firestone RB, Tiedje JM (1980) Nitrous oxide from soil denitrification: factors controlling its biological production. Science 208:749–751.

Firestone MK (1982) Biological denitrification. In: Stevenson FG (ed). Nitrogen in Agricultural Soils. American Society of Agronomy, Madison, WI, pp 289–326.

Firestone MK, McColl JG, Killham KS, Brooks PD (1984) Microbial response to acid deposition and effects on plant productivity. In: Linhurst RA (ed) Direct and Indirect Effects of Acidic Deposition on Vegetation. Ann Arbor Science Publishing, Ann Arbor, MI, pp 51–63.

Fischer P, Führer E (1990) Effect of soil acidity on the entomophilic nematode *Steinernema kraussei* Steiner. Biol Fertil Soils 9:174–177.

Fowler D (1984) Transfer to terrestrial surfaces. Philos Trans R Soc London B Biol Sci 305:281–297.

Francis AJ, Olson D, Bernatsky R (1980) Effects of acidity on microbial processes in a forest soil. Presented at the International Conference on the Ecological Impact of Acid Precipitation, ASNLH, Norway, 11–14 March 1980. BNL-27848. Brookhaven National Laboratory, Upton, NY.

Francis AJ (1982) Effects of acidic precipitation and acidity on soil microbial processes. Water Air Soil Pollut 18:375–394.

Franz H (1959) Das diologische Geschehen im Waldboden und seine Beeinflussung durch die Kalkdungung. Allg Forst Ztg 70:178–181.

Freedman B, Hutchinson TC (1980) Smelter pollution near Sudbury, Ontario, Canada, and effects on forest litter decomposition. In: Hutchinson TC, Havas M (eds) Effects of Acidic Precipitation on Terrestrial Ecosystems. NATO ASI Ser 1 Global Environ Change 4:395–434.

French DD (1988) Some effects of changing soil chemistry on decomposition of plant litters and cellulose on a Scottish moor. Oecologia 75:608–618.

Fritze H, Kiikkilä O, Pasanen J, Pietikäinen J (1992) Reaction of forest soil microflora to environmental stress along a moderate pollution gradient next to an oil refinery. Plant Soil 140:175–182.

Furlan V, Bernier-Cardou M (1989) Effects of N, P, and K on formation of vesicular-arbuscular mycorrhizae, growth and mineral content of onion. Plant Soil 113:167–174.

Gagnon J, Langlois CG, Garbaye J (1991) Growth and ectomycorrhiza formation of container-grown red oak seedlings as a function of nitrogen fertilization and inoculum type of *Laccaria bicolor*. Can J For Res 21:966–973.

Galloway JN, Likens GE, Hawley ME (1984) Acidic precipitation: natural versus anthropogenic components. Science 226:829–831.

Garbaye J, Kabre A, Le Tacon F, Mousain D, Piou D (1979) Fertilization minérale et fructification des champignons supérieurs en hêtraie. Ann Sci For (Paris) 36: 151–164.

Garbaye J, Le Tacon F (1982) Influence of mineral fertilization and thinning intensity on the fruit body production of epigeous fungi in an artificial spruce stand (*Picea excelsa* Link) in north-eastern France. Acta Oecol Oecol Plant 3:153–160.

Garden A, Davies RW (1988) The effects of a simulated acid precipitation on leaf litter quality and the growth of a detritivore in a buffered lotic system. Environ Pollut 52:303–313.

Gärdenfors U (1992) Effects of artificial liming on land snail populations. J Appl Ecol 29:50–54.

Garland JA (1977) The dry deposition of sulfur dioxide to land and water surfaces. Proc R Soc London A 354:245–268.

Gates CE (1978) Contributions to a revision of the earthworm family Lumbricidae. XXII. The genus *Eisenia* in North America. Megadrilogica 3:131–147.

Gerard BM, Hay RKM (1979) The effect on earthworms of ploughing, tined cultivation, direct drilling and nitrogen in a barley monoculture system. J Agric Sci 93: 147–155.

Germida JJ, Wainwright M, Gupta VVSR (1992) Biochemistry of sulfur cycling in soil. In: Stotzky G, Bollag J-M (eds) Soil Biochemistry. Marcel Dekker, New York, pp 1–53.

Ghani A, McLaren RG, Swift RS (1992) Sulphur mineralization and transformations in soils as influenced by additions of carbon, nitrogen and sulphur. Soil Biol Biochem 24:331–341.

Ghiorse WC, Alexander M (1976) Effects of microorganisms on the sorption and fate of sulfur dioxide and nitrogen dioxide in soil. J Environ Qual 5:227–230.

Göbl F (1988) Mykorrhiza-und Feinwurzeluntersuchungen im Waldschadensgebiet Gleingraben/Steiermark. Österr Forsztztg 6:16–18.

Gorham E (1955) On the acidity and salinity of rain. Geochim Cosmochim Acta 7: 231–239.

Gorissen A, Joosten NN, Burgers SLGE (1994) Ammonium deposition and the mycoflora in the rhizosphere of Douglas-fir. Soil Biol Biochem 26:1011–1022.

Gosz JR (1981) Nitrogen cycling in coniferous ecosystems. In: Clark FE, Rosswall T (eds) Terrestrial Nitrogen Cycles. Ecol Bull 33:405–426.

Gould RP, Minchin PEH, Young PC (1988) The effects of sulfur dioxide on phloem transport in two cereals. J Exp Bot 39:997–1007.

Granhall U, Selander H (1973) Nitrogen fixation in subarctic mire. Oikos 24:8–15.

Grant IF, Bancroft K, Alexander M (1979) SO_2 and NO_2 effects on microbial activity in acid forest soil. Microb Ecol 5:85–89.

Gray TRG, Ineson P (1981) The effects of sulphur dioxide and acid rain on the decomposition of leaf litter. J Sci Food Agric 32:624–625.

Greszta J, Gruszka A, Wachalewski T (1992) Humus degradation under the influence of simulated 'acid rain'. Water Air Soil Pollut 63:51–66.

Gundersen P (1991) Nitrogen deposition and the forest nitrogen cycle: role of denitrification. For Ecol Manage 44:15-28.

Gunnarsson T, Rundgren S (1986) Nematode infestation and hatching failure of lumbricid cocoons in acidified and metal polluted soils. Pedobiologia 29:165-173.

Gupta VVSR, Jawrence JR, Germida JJ (1988) Impact of elemental sulfur fertilization on agricultural soils. I. Effects on microbial biomass and enzyme activities. Can J Soil Sci 68:463-473.

Hågvar S, Abrahamsen G (1977a) Effect of artificial acid rain on Enchytraeidae, Collembola and Acarina in coniferous forest soil, and on Enchytraeidae in sphagnum bog—preliminary results. In: Lohm U, Persson T (eds) Soil Organisms as Components of Ecosystems. Proceedings of the VI International Soil and Zoology Colloquium, Stockholm. Ecol Bull 25:568-570.

Hågvar S, Abrahamsen G (1977b) Acidification experiments in conifer forest. 5. Studies on the soil fauna. In: Acid Precipitation—Effects on Forest and Fish—Project Norway. Internal Rep 32:1-47.

Hågvar S (1978) Acidification experiments in coniferous forests. 6. Effects of acidification and liming on Collembola and Acarina. SNSF,Oslo, Norway. IR36/78.

Hågvar S (1980) Effects of artificial acid precipitation on soil and forests. 7. Soil animals. p 202-203. In: Drabloes D, Tollan A (eds) Ecological Impact of Acid Precipitation. Proceedings of an International Conference. SNSF, Oslo, Norway.

Hågvar S, Abrahamsen G (1980) Colonization by Enchytraeidae, Collembola and Acari in sterile soil samples with adjusted pH levels. Oikos 34:245-258.

Hågvar S, Amundsen T (1981) Effect of liming and artificial acidrain on the mite (*Acari*) fauna in coniferous forest. Oikos 37:7-20.

Hågvar S, Kjøndal BR (1981a) Decomposition of birch leaves: dry weight loss, chemical changes, and effects of artificial acid rain. Pedobiologia 22:232-245.

Hågvar S, Kjøndal BR (1981b) Effects of artificial acid rain on the microarthropod fauna in decomposing birch leaves. Pedobiologia 22:409-422.

Hågvar S (1984a) Effects of liming and artificial acid rain on Collembola and Protura in coniferous forest. Pedobiologia 27:341-354.

Hågvar S (1984b) Six common mite species (Acari) in Norwegian coniferous forest soils: relations to vegetation types and soil characteristics. Pedobiologia 27:355-364.

Hågvar S, Abrahamsen G (1984) Collembola in Norwegian coniferous forest soils. III. Relation to soil chemistry. Pedobiologia 27:331-339.

Hågvar S (1987a) Effects of artificial acid precipitation and liming on forest microarthropods. In: Proceedings of the 9th International Colloquium on Soil Zoology, Moscow. Nauka, USSR, pp 661-668.

Hågvar S (1987b) What is the importance of soil acidity for the soil fauna? Fauna 40:64-72.

Hågvar S (1988a) Acid rain and soil fauna. In: Iturrondobeitia JC (ed) Biologia Ambiental. Tomo I. Proceedings of the Second World Basque Congress, Bilbao, November 1987. Universidad del Pais Vasco, Bilbao, pp 191-201.

Hågvar S (1988b) Decomposition studies in an easily-constructed microcosm: effects of microarthropods and varying soil pH. Pedobiologia 31:293-303.

Hågvar S (1990) Reactions to soil acidification in microarthropods: is competition a key factor? Biol Fertil Soils 9:178-181.

Halstead RL (1964) Phosphate activity of soils as influenced by lime and other treatments. Can J Soil Sci 44:137–144.

Hancock JG, Millar RL, Lorbeer JW (1964) Pectolytic and cellulolytic enzymes produced by *Botrytis allii, B. cinerea* and *B. squamosa* in vitro and in vivo. Phytopathology 54:928–931.

Harley JL, Smith SE (1983) Mycorrhizal Symbiosis. Academic Press, London.

Harrison AF (1987) Soil Organic Phosphorus. CAB International, Wallingford.

Hartenstein R (1962) Soil Oribatei. I. Feeding specificity among forest soil Oribatei (Acarina). Ann Entomol Soc Am 55:202–206.

Hauhs M, Rost-Siebert K, Raben G, Paces T, Vigerus B (1989) Summary of European data. In: Malanchuk JL, Nilsson J (eds). The Role of Nitrogen in the Acidification of Soils and Surface Waters. Nordic Council of Ministers, Copenhagen, Denmark.

Hayman DS (1970) Endogone spore numbers in soil and vesicular arbuscular mycorrhiza in wheat as influenced by season and soil treatment. Trans Br Mycol Soc 54:53–63.

Hayman DS, Travares M (1985) Plant growth responses to vesicular-arbuscular mycorrhiza. XV. Influence of soil pH on the symbiotic efficiency of different endophytes. New Phytol 100:367–377.

Haynes RJ, Swift RS (1988) Effects of lime and phosphate additions on changes in enzyme activities, microbial biomass and levels of extractable nitrogen, sulphur and phosphorus in an acid soil. Biol Fertil Soil 6:153–158.

Heijne B, van Dam D, Heil GB, Bobbink R (1989a) The influence of the "acid rain" component ammonium sulphate on vesicular arbuscular mycorrhiza. In: Brasser LJ, Mulder WC (eds) Man and His Ecosystem. Proceedings of the 8th World Clean Air Congress 1989, The Hague, Vol. 2. Elsevier, The Netherlands, pp 257–261.

Heijne B, Heil GB, van Dam D (1989b) Relations between acid rain and vesicular-arbuscular mycorrhiza. Agric Ecosyst Environ 29:187–192.

Heilman PE (1975) Effects of added salts on nitrogen release and nitrate levels in forest soils of the Washington coastal area. Soil Sci Soc Am Proc 39:778–782.

Hendrickson OQ (1985) Variation in the C : N ratio of substrate mineralized during forest humus decomposition. Soil Biol Biochem 17:435–440.

Hepper CM (1983) The effect of nitrate and phosphate on the vesicular arbuscular mycorrhizal infection of lettuce. New Phytol 93:389–399.

Hepper CM (1984) Inorganic sulfur nutrition of the vesicular-arbuscular mycorrhizal fungus *Glomus caledonium*. Soil Biol Biochem 16:669–671.

Hern JA, Rutherford GK, vanLoon GW (1985) Chemical and pedogenetic effects of simulated acid precipitation on two eastern Canadian forest soils. I. Nonmetals. Can J For Res 15:839–847.

Heungens A (1981) Nematode populations fluctuations in a pine litter after treatment with pH changing compounds. Meded Fac Landbouww Rijksuniv Gent 46:1267–1281.

Heungens A, van Daele E (1984) The influence of some acids, bases and salts on the mite and Collembola population of a pine litter substrate. Pedobiologia 27:299–311.

Hile N, Hennen JF (1969) *In vitro* culture of *Pisolithus tinctorius* mycelium. Micologia 61:195–198.

Ho I, Trappe JM (1984) Effects of ozone exposure on mycorrhiza formation and growth of *Festuca arundinacea*. Environ Exp Bot 24:71–74.

Homann PS, Cole DW (1990) Sulfur dynamics in decomposing forest litter: relationship to initial concentration, ambient sulfate and nitrogen. Soil Biol Biochem 22:621-628.

Hovland J, Abrahamsen G, Ogner G (1980) Effects of artificial acid rain on decomposition of spruce needles and on mobilization and leaching of elements. Plant Soil 56:365-378.

Hovland J (1981) The effect of artificial acid rain on respiration and cellulase activity in Norway spruce needle letter. Soil Biol Biochem 13:23-26.

Hryniuk J (1966) Influence of many years fertilization on the mesofauna in soil. In: Rapoport EH (ed) Progresos en Biologia del Suelo. Ler Coloquio Latinoamericano de Biologia del Suelo, Bahia Blance, 1965, Montevideo, pp 413-417.

Huhta V (1984) Response of *Cognettia sphagnetorum* (Enchytraeidae) to manipulation of pH and nutrient status in coniferous forest soil. Pedobiologia 27:254-260.

Huhta V, Karppinen E, Nurminen M, Valpas A (1967) Effect of silvicultural practices upon arthropod, annelid and nematode populations in coniferous forest soil. Ann Zool Fenn 4:87-143.

Huhta V, Matli N, Valpas A (1969) Further notes on the effect of silvicultural practices upon the fauna of coniferous forest soil. Ann Zool Fenn 6:327-334.

Huhta V, Hyvönen R, Koskenniemi A, Vilkamaa P (1983) Role of pH in the effect of fertilization on Nematoda, Oligochaeta and Microarthropoda. In: Lebrun P, Andre HM, Medts AD, Gregoire-Wibo C, Wauthy G (eds) New Trends in Soil Biology. Proceedings VIII. International Colloquium of Soil and Zoology, Louvain-la-Neuve, pp 61-73.

Huhta V, Hyvönen R, Koskenniemi A, Vilkamaa P, Kaasalainen P, Sulander M (1986) Response of soil fauna to fertilization and manipulation of pH in coniferous forests. Acta For Fenn 195:1-30.

Hung LL, Trappe JM (1983) Growth variation between and within species of ectomycorrhizal fungi in response to pH *in vitro*. Mycologia 75:234-241.

Huther W (1959) Zur ernahrung der Pauropoden. Naturwissenschaften 19:563-564.

Hutson BR (1978) Influence of pH, temperature and salinity on the fecundity and longevity of four species of Collembola. Pedobiologia 18:163-179.

Hyvönen R, Huhta V (1989) Effects of lime, ash and nitrogen fertilizers on nematode populations in Scots pine forests. Pedobiologia 33:129-143.

Hyvönen R, Persson T (1990) Effects of acidification and liming on feeding groups of nematodes in coniferous forest soils. Biol Fertil Soils 3:57-68.

Illmer P, Schinner F (1991) Effects of lime and nutrient salts on the microbiological activities of forest soils. Biol Fertil Soils 11:261-266.

Illmer P, Marschall K, Schinner F (1995) Influence of available aluminum on soil microorganisms. Lett Appl Microbiol 21:393-397.

Ineson P (1983) The effect of airborne sulphur pollutants upon decomposition and nutrient release in forest soils. Ph.D. thesis, University of Liverpool, UK.

Jarvis BW, Land GE, Wieder RK (1987) Arylsulfatase activity in peat exposed to acid precipitation. Soil Biol Biochem 19:107-109.

Jefferson P (1955) Studies on earthworms of turf. B. Earthworms and soil. J Sports Turf Res Inst 9:166-179.

Johnson AC, Wood M (1990) DNA, a possible site of action of aluminum in *Rhizobium* spp. Appl Environ Microbiol 56:3629-3633.

Johnson AH, Siccama TJ (1983) Acid deposition and forest decline. Environ Sci Technol 17:294-306.

Johnson DD, Guenzi WD (1963) Influence of salts on ammonium oxidation and carbon dioxide evolution from soil. Soil Sci Soc Am Proc 27:663-666.

Johnson DW, Hornbeck JW, Kelly JM, Swank WT, Todd DE (1980) Regional pattern of soil sulfate accumulation: relevance to ecosystem sulfur budgets. In: Shriner DS, Richmond CR, Lindberg SE (eds) Atmospheric Sulfur Deposition. Environmental Impact and Health Effects. Ann Arbor Press, Ann Arbor, MI, pp 507-520.

Johnson DW, Henderson GS, Huff DD, Lindberg SE, Richter DD, Shriner DS, Todd DE, Turner J (1982) Cycling of organic and inorganic sulfur in a chestnut oak forest. Oecologia 54:141-148.

Johnson DW (1984) Sulfur cycling in forests. Biogeochemistry (Dordrecht) 1:29-43.

Johnson DW, Todd DE (1984) Effects of acid irrigation on carbon dioxide evolution, extractable nitrogen, phosphorus, and aluminum in deciduous forest soil. Soil Sci Soc Am J 48:664-666.

Johnson DW, Kelly JM, Swank WT, Cole DW, Hornbeck JW, Pierce RS, Van Lear D (1985) A comparative evaluation of the effects of acid precipitation, natural acid production, and harvesting on cation removal from forests. ERD Publ 2508. Oak Ridge National Laboratory, Oak Ridge, TN.

Johnson DW, Van Miegroet H, Kelly JM (1986) Sulfur cycling in five forest ecosystems. Water Air Soil Pollut 30:965-979

Johnson DW (1987) Acid deposition and forest nutrient cycling. ESD Rep. 870416. Oak Ridge National Laboratory, Oak Ridge, TN, pp 1-13.

Kabata-Pendias A, Pendias H (1986) Trace Elements in Soils and Plants. CRC Press, Boca Raton, FL.

Kaplan DL, Hartenstein R, Neuhauser EF, Malecki MR (1980) Physicochemical requirements in the environment of the earthworm *Eisenia foetida*. Soil Biol Biochem 12:347-352.

Kardell L, Eriksson L (1987) Kremlor, riskor, soppar. Skogsbruksmetodernas inverkan på productionen av matsvamar. Sver Skogsvårdsförb Tidskr 2:3-23.

Keane KD, Manning WJ (1988) Effects of ozone and simulated acid rain on birch seedlings growth and formation of ectomycorrhizae. Environ Pollut 52:55-56.

Kelly JM, Strickland RC (1984) CO_2 efflux from deciduous forest litter and soil in response to simulated acid rain treatment. Water Air Soil Pollut 23:431-440.

Killham K, Wainwright M (1981) Deciduous leaf litter and cellulose decomposition in soil exposed to heavy atmospheric pollution. Environ Pollut Ser A Ecol Biol 26:79-85.

Killham K, Firestone MK (1982) Evaluation of accelerated H_+ applications in predicting soil chemical and microbial changes due to acid rain. Commun Soil Sci Plant Anal 13:995-1001.

Killham K, Firestone MK (1983) Vesicular-arbuscular mycorrhizal mediation of grass response to acidic and heavy metal depositions. Plant Soil 72:39-48.

Killham K, Firestone MK, McColl JC (1983) Acid rain and soil microbial activity: effects and their mechanisms. J Environ Qual 12:133-137.

Klein TM, Novick NJ, Kreitinger SP, Alexander M (1984) Simultaneous inhibition of carbon and nitrogen mineralization in forest soil by simulated acid precipitation. Bull Environ Contam Toxicol 32:698-703.

Koskenniemi A, Huhta V (1986) Effects of fertilization and manipulation of pH on mite (Acari) populations of coniferous forest soil. Rev Ecol Biol Sol 23:271-286.

Kratz W, Brose A, Weigmann G (1991) The influence of lime application in damaged pine forest ecosystems in Berlin (FRG): soil chemical and biological aspects. In: Ravera O (ed) Terrestrial and Aquatic Ecosystems: Perturbation and Recovery. Ellis Horwood, Chichester, pp 464–471.

Kreutzer K (1995) Effects of forest liming on soil processes. Plant Soil 168-169:447–470.

Kuhnelt W (1961) Soil Biology with Special Reference to the Animal Kingdom. Faber and Faber, London.

Kuperman R (1995) Abundance of soil macroinvertebrates in oak-hickory forests along the Ohio river acidic deposition gradient. Acta Zool Fenn 196:76–79.

Kuperman R (1996) Relationships between soil properties and community structure of soil macroinvertebrates in oak-hickory forests along an acidic deposition gradient. Appl Soil Ecol (in press).

Ladd JN (1978) Origin and Range of Enzymes in Soil. In: Burns RG (ed) Soil Enzymes. Academic Press, London, pp 51–96.

Lampky JR, Peterson JE (1963) *Pisolithus tinctorius* associated with pines in Missouri. Micologia 55:675–678.

Lang E, Beese F (1985) Die Reaktion der microbiellen Bodenpopulation eines Buchenwaldes auf Kalkungsmaßnahmen. Allg Forstztg 43:1166–1169.

Langkramer O, Lettl A (1982) Influence of industrial atmospheric pollution on soil biotic component of Norway spruce stands. Zentralbl Mikrobiol 137:180.

Larkin RP, Kelly JM (1987) Influence of elevated ecosystem S levels on litter decomposition and mineralization. Water Air Soil Pollut 34:415–428.

Larkin RP, Kelly JM (1988a) A short-term microcosm evaluation of CO_2 evolution from litter and soil as influenced by SO_2 and SO_4 additions. Water Air Soil Pollut 37:273–280.

Larkin RP, Kelly JM (1988b) Influence of elevated ecosystem S levels on litter decomposition and mineralization. Water Air Soil Pollut 34:415–428.

Last FT, Likens GE, Ulrich B, Walloe L (1980) Acid precipitation — progress and problems. Conference summary. In: Drablos D Tollan A (eds) Ecological Impact of Acid Precipitation. SNSF, Oslo, Norway, pp 10–12.

Lee JJ, Weber DE (1983) Effects of sulfuric acid rain on decomposition rate and chemical element content of hardwood leaf litter. Can J Bot 61:872–879.

Leetham JW, McNary TJ, Dodd JL, Laurenroth WK (1980) Response of field populations of Tardigrada to various levels of chronic low-level sulphur dioxide exposure. In: Dindal DL (ed) Soil Biology as Related to Land Use Practices. U.S. Environmental Protection Agency, Washington, DC, pp 382–390.

Leetham JW, McNary TJ, Dodd JL, Laurenroth WK (1982) Response of soil nematodes, rotifers and tardigrades to three levels of season-long sulfur dioxide exposures. Water Air Soil Pollut 17:343–356.

Leetham JW, Dodd JL, Laurenroth WK (1983) Effects of low-level sulfur dioxide exposure on decomposition of *Agropyron smithii* litter under laboratory conditions. Water Air Soil Pollut 19:247–250.

Lehto T (1994) Effects of soil pH and calcium on mycorrhizas of *Picea abies*. Plant Soil 163:69–75.

Leone G, Van den Heuvel J (1986) Regulation by carbohydrates of the sequential *in vitro* production of pectic enzymes by *Botrytus cinerea*. Can J Bot 65:2133–2141.

Lettl A (1981a) Effect of some sulfur compounds on soil microflora of spruce rhizosphere. Folia Microbiol 26:243–252.

Lettl A (1981b) The effect of emissions on microbiology of the sulphur cycle. Commun Inst For Czech 12:27–50.

Lettl A (1984) The effect of atmospheric SO_2 pollution on the microflora of forest soils. Folia Microbiol 29:455–475.

Lettl A (1986) Biochemical activities of soil microflora in SO_2 polluted forest stands. Folia Microbiol 31:220–227.

Levine ER, Ciolkosz EJ (1988) Computer simulation of soil sensitivity to acid rain. Soil Sci Soc Am J 52:209–215.

Likens GE, Borman FH (1974) Acid rain: a serious regional environmental problem. Science 184:1176–1179.

Likens GE, Borman FH, Pierce AS, Eaton JS, Johnson NM (1977) Biogeochemistry of a Forested ecosystem. Springer-Verlag, New York.

Lisker N, Katan J, Henis Y (1975) Sequential production of polygalacturonase, cellulase, and pectin lyase by *Rhizoctonia solani*. Can J Microbiol 21:1298–1304.

Ljungholm K, Norén B, Wadsö I (1979) Microcalorimetric observations of microbial activity in normal and acidified soils. Oikos 33:24–30.

Lohm U, Lundkvist H, Persson T, Wiren A (1977) Effects of nitrogen fertilization on the abundance of enchytraeids and microarthropods in Scots pine forests. Stud For Suec 140:1–23.

Lohm U (1980) Effects of experimental acidification on soil organism populations and decomposition. In: Drabloes D, Tollan A (eds) Ecological Impact of Acid Precipitation. Proceedings of an international conference. SNSF, Oslo, Norway, pp 178–179.

Lohm U, Larsson K, Nömmik H (1984) Acidification and liming of coniferous forest soil: long-term effects on turnover rates of carbon and nitrogen during an incubation experiment. Soil Biol Biochem 16:343–346.

Loucks OL, Kuperman R (1991) Effects of drought stress on soil invertebrate communities in oak-hickory forests of the Ohio Corridor Pollution gradient. Ecological Society of America meetings, San Antonio, Texas, August 4–8, 1991.

Loucks OL (1992) Forest response research in NAPAP: potentially successful linkage of policy and science. Ecol Appl 2:117–123.

Lundkvist H (1977) Effects of artificial acidification on the abundance of Enchytraeidae in a Scots pine forest in northern Sweden. Soil organisms as components of ecosystems. Ecol Bull (Stockholm) 25:570–572.

Ma W-C (1982) The influence of soil properties and worm-related factors on the concentration of heavy metals in earthworms. Pedobiologia 24:109–119.

Ma W-C, Edelman T, Beersum I van, Jans T (1983) Uptake of cadmium, zinc, lead, and copper by earthworms near a zinc-smelting complex: influence of soil pH and organic matter. Bull Environ Contam Toxicol 30:424–427.

Ma W-C, Brussaard L, de Ridder JA (1990) Long-term effects of nitrogenous fertilizers on grassland earthworms (Oligochaeta: Lumbricidae): their relation to soil acidification. Agric Ecosyst Environ 30:71–80.

Maccari G, Ciardi C, Ceccanti B, Masciandaro G (1994) Biochemical study in a microcosm of a soil exposed to high levels of SO_2. Geomicrobiol J 11:317–323.

Maclagan DS (1932) An ecological study on the "Lucerne flea" (*Sminthurus viridis* L.). Bull Entomol Res 23:101–145.

Mai H, Fiedler HJ (1989) Model and field trials on the effect of SO_2 on microorganisms in spruce raw humus. Zentralbl Mikrobiol 144:129–136.

Mai H (1990) Soil-microbial investigations carried out in the ecological monitoring station in Tharandt Forest. Zentralbl Mikrobiol 145:293–304.

Mai H, Fiedler HJ (1990) The effect of artificial fumigation with SO$_2$ on microorganisms in limed and unlimed raw humus from a spruce stand. Zentralbl Mikrobiol 145:157–163.

Mancinelli RL (1986) Alpine tundra soil bacterial responses to increased soil loading rates of acid precipitation, nitrate, and sulfate. Front Range, Colorado, USA. Arct Alp Res 18:269–275.

Marschner B, Wilczynski AW (1991) The effect of liming on quantity and chemical composition of soil organic matter in a pine forest in Berlin, Germany. Plant Soil 137:229–236.

Marshall VG (1974) Seasonal and vertical distribution of soil fauna in a thinned and urea-fertilized Douglas fir forest. Can J Soil Sci 54:491–500.

Marshall VG (1977) Effects of manures and fertilizers on soil fauna: a review. Commonwealth Agricultural Bureau, England.

Martikainen PJ (1985) Nitrification in forest soil of different pH as affected by urea, ammonium sulphate, and potassium sulphate. Soil Biol Biochem 17:363–367.

Martikainen PJ, Aarnio T, Taavitsainen V-M, Päivinen L, Salonen K (1989) Mineralization of carbon and nitrogen in soil samples taken from three fertilized pine stands: long-term effects. Plant Soil 114:99–106.

Martikainen PJ, Aarnio T, Yrjälä K (1990) Long-term effects of nitrogen additions on mineralization of carbon and nitrogen in forest soils. In: Brandon O, Hüttl RF (ed) Nitrogen Saturation in Forest Ecosystems. Kluwer, The Netherlands, pp 121–126.

McAfee BJ, Fortin JA (1987) The influence of pH on the competitive interactions of ectomycorrhizal mycobionts under field conditions. Can J For Res 17:859–864.

McAndrew DW, Malhi SS (1992) Long-term N fertilization of a Solonetzic soil: effects on chemical and biological properties. Soil Biol Biochem 24:619–623.

McBrayer JF, Reichle DE, Auerbach SI (1970) Trophic level delineation of forest microinvertebrates. Tech manual #2847. Oak Ridge National Laboratory, Oak Ridge, TN.

McColl JG, Firestone MK (1987) Cumulative effects of simulated acid rain on soil chemical and microbial characteristics and conifer seedling growth. Soil Sci Soc Am J 51:794–800.

McColl JG, Firestone MK (1991) Soil chemical and microbial effects of simulated acid rain on clover and chess. Water Air Soil Pollut 60:301–313.

McCool PM, Menge JA, Taylor OC (1979) Effects of ozone and HCl gas on the development of the mycorrhizal fungus *Glomus fasciculatus* and growth of 'Troyer' citrage. J Am Soc Hortic Sci 104:151–154.

McGill WB, Hunt HW, Woodmansee RG, Reuss JO (1981) Ecol Bull (Stockholm) 33:49.

McLeod AR (1988) Effects of open-air fumigation with sulphur dioxide on the occurrence of fungal pathogens in winter cereals. Phytopathology 78:88–94.

McQuattie CJ, Schier GA (1987) Effects of ozone and aluminum on pitch pine ectomycorrhizae. In: Sylvia DM, Hung LL, Graham (eds) Mycorrhizae in the Next Decade. Proceedings of the 7th NACOM, Gainesville, FL.

Meier S, Robarge WP, Bruck RI, Grand LF (1989) Effects of simulated rain acidity on mycorrhizae of red spruce seedlings potted in natural soils. Environ Pollut 59:314–315.

Meiwes KJ, Khanna PK (1981) Distribution and cycling of sulphur in the vegetation of two forest ecosystems in an acid rain environment. Plant Soil 60:369–375.

Melillo JM (1981) Nitrogen cycling in deciduous forests. In: Clark FE, Rosswall T (eds) Terrestrial Nitrogen Cycles. Ecol Bull 33:427–442.

Mersi W von, Kuhnert-Finkernagel R, Schinner F (1992) The influence of rock powders on microbial activity of three forest soils. Z Pflanzenernähr Dueng Bodenkd 155:29–33.

Mikola P (1973) Application of mycorrhizal symbiosis in forestry practice. In: Marks GC, Kozlowski TT (eds) Ectomycorrhizae: Their Ecology and Physiology. Academic Press, New York, pp 383–411.

Miller KW, Cole MA, Banwart WL (1991) Microbial populations in an agronomically managed mollisol treated with simulated acid rain. J Environ Qual 20: 845–849.

Miller RM (1987) The ecology of vesicular-arbuscular mycorrhizae in grass- and shrublands. In: Safir GR (ed) Ecophysiology of VA Mycorrhizal Plants, CRC Press, Boca Raton, FL, pp 135–170.

Mladenoff DJ (1987) Dynamics of nitrogen mineralization and nitrification in hemlock and hardwood treefall gaps. Ecology 68:1171–1180.

Modaihsh AS, Al-Mustafa WA, Metwally AI (1989) Effect of elemental sulphur on chemical changes and nutrient availability in calcareous soils. Plant Soil 116:95–101.

Mohren GMJ, Van Den Burg J, Burger FW (1986) Phosphorus deficiency induced by nitrogen input in Douglas fir in the Netherlands. Plant Soil 95:191–200.

Moloney KA, Stratton LJ, Klein RM (1983) Effects of simulated acidic, metal-containing precipitation on coniferous litter decomposition. Can J Bot 61:337–342.

Moore TR (1987) The effect of simulated acid rain on the nutrient status of subarctic woodland soils in eastern Canada. Can J For Res 17:370–378.

Mrkva R, Grunda B (1969) Einfluss von Immissionen auf die Waldboden und ihre Mikroflora im Gebiet von Sudmahren. Acta Univ Agric Fac Silvic (Brno) 38:247.

Mulder EG, Brotonegoro S (1974) Free-living heterotrophic nitrogen-fixing bacteria. In: Quispel A (ed) The Biology of Nitrogen Fixation. Plenum Press, Amsterdam, pp 342–378.

Munger JW, Eisenrich SJ (1983) Continental scale variations in precipitation chemistry. Environ Sci Technol 17:32A–42A.

Munns DH, Keyser HH (1981) Responses of *Rhizobium* strains to acid and aluminium stress. Soil Biol Biochem 13:115–118.

Myrold DD (1987) Effects of acidic deposition on soil organisms. In: Acidic deposition and forest soil biology. Tech Bull 527. NCASI, New York, pp 1–29.

Myrold DD (1990) Effects of acidic deposition on soil organisms. In: Lucier AA, Haines SG (eds) Mechanisms of Forest Response to Acidic Deposition. Springer-Verlag, New York, pp 163–187.

Myrold DD, Nason GE (1992) Effect of Acid Rain on Soil Microbial Processes. Environmental Microbiology. Wiley-Liss, New York, pp 59–81.

Nahas E, Terenzi HF, Rossi A (1982) Effect of carbon source and pH on the production and secretion of acid phosphatase (EC.3.1.3.2) and alkaline phosphatase (EC.3.1.3.1) in *Neurospora crassa*. J Gen Microbiol 128:2017–2021.

National Acid Precipiation Assessment Program (NAPAP) (1987) Interim assessment: the causes and effect of acidic deposition. Vol. 1: Executive Summary. National Acid Precipitation Assessment Program, CEQ, Washington, DC.

NAPAP (1990) Deposition monitoring: methods and results. In: Sisterson DS, ed. Acidic Deposition: State of Science and Technology. NAPAP Rep 6, Washington, DC.

Neuvonen S, Suomela J (1990) The effects of simulated acid rain on pine needle and birch leaf litter decomposition. J Appl Ecol 27:857–872.

Newsham KK, Frankland JC, Boddy L, Ineson P (1992a) Effects of dry-deposited sulfur dioxide on fungal decomposition of angiosperm tree leaf litter. I. Changes in communities of fungal saprotrophs. New Phytol 122:97–110.

Newsham KK, Frankland JC, Boddy L, Ineson P (1992b) Effects of dry-deposited sulfur dioxide on fungal decomposition of angiosperm tree leaf litter. II. Chemical content of leaf litters. New Phytol 122:111–125.

Newsham KK, Frankland JC, Boddy L, Ineson P (1992c) Effects of dry-deposited sulfur dioxide on fungal decomposition of angiosperm tree leaf litter. III. Decomposition rates and fungal respiration. New Phytol 122:127–140.

Newsham KK, Ineson P, Frankland JC (1995) The effects of open-air fumigation with sulfur dioxide on the decomposition of sycamore (*Acer preudoplatanus* L.) leaf litters from polluted and unpolluted woodlands. Plant Cell Environ 18:309–319.

Nilsson J, Grennfelt P (eds) (1988) Critical loads of sulphur and nitrogen. Nordic Council of Ministers, Copenhagen.

Nilsson SI, Miller HG, Miller JD (1982) Forest growth as a possible cause of soil and water acidification: an examination of the concepts. Oikos 39:40–49.

Nodar R, Acea MJ, Carballas T (1992) Microbial response to $Ca(OH)_2$ treatments in a forest soil. FEMS Microbiol Ecol 86:213–219.

Nohrstedt H-Ö (1985) Studies of forest floor biological activities in an area previously damaged by sulphur dioxide emissions. Water Air Soil Pollut 25:301–311.

Nohrstedt H-Ö (1988) Effects of liming and N-fertilization on denitrification and N_2-fixation in an acid coniferous forest floor. For Ecol Manage 24:1–13.

Nohrstedt H-Ö, Arnebrant K, Bååth E, Söderström B (1989) Changes in carbon content, respiration rate, ATP content, and microbial biomass in nitrogen-fertilized pine forest soils in Sweden. Can J For Res 19:323–328.

Noordwijk MV, Hairiah K (1986) Mycorrhizal infection in relation to soil pH and soil phosphorus content in a rain forest of northern Sumatra. Plant Soil 96:299–302.

Nylund J-E (1988) The regulation of mycorrhiza formation — carbohydrate and hormone theories reviewed. Scand J For Res 3:465–470.

Oelbe-Farivar M (1985) Physiologishe Reaktionen von Mykorrhizapilzen auf simulierte saure Bodenbedingungen. Ph.D thesis, University of Gottingen.

Olson RA (1983) The impacts of acid deposition on N and S cycling. Environ Exp Bot 23:211–223.

Overrein L, Seip H, Tollan A (1981) Acid precipitation effects on forest and fish. Final report of the SNSF project 1972–1980. SNSF, Oslo, Norway.

Padan E (1984) Adaptation of bacteria to external pH. In: Klug MJ, Reddy CA (eds) Current Perspectives in Microbial Ecology. American Society for Microbiology, Washington, DC, pp 49–54.

Persley AF, Page OT (1971) Differential induction of pectolytic enzymes of *Fusarium roseum* (Lk.) emend. Snyder and Hanses. Can J Microbiol 17:415–420.

Persson T (1983) In: Lebrun PH, et al. (eds) New Trends in Soil Biology. Proceedings, VIII International Colloquium on Soil and Zoology. p 117.

Persson T, Lundkvist H, Wirén A, Hyvönen R, Wessén B (1989) Effects of acidification and liming on carbon and nitrogen mineralization and soil organisms in moor humus. Water Air Soil Pollut 45:77-96.

Phaff HJ (1947) The production of exocellular pectic enzymes by *penicillium chrysogenum*. I. On the formation and adaptive nature of polygalacturonase and pectin esterase. Arch Biochem 13:67-81.

Placet M (1991) Emissions involved in acidic deposition processes. Rep 1. NAPAP State of Science and Technology. NAPAP, Washington, DC.

Popovic B (1984) Mineralization of carbon and nitrogen in humus from field acidification studies. For Ecol Manage 8:31-93.

Postel S (1984) Air pollution, acid rain, and the future of forests. Paper #58. Worldwatch Institute, Washington, DC.

Potter DA, Bridges BL, Gordon FC (1985) Effect of N fertilization on earthworm and microarthropod populations in Kentucky bluegrass turf. Agron J 77:367-372.

Prescott CE, Bewley RJF, Parkinson D (1984) Litter decomposition and soil microbial activity in a forest receiving SO_2 pollution. In: Stone EL (ed) Forest Soils and Treatment Impacts. University of Tennessee, Knoxville, p 448.

Prescott CE, Parkinson D (1985) Effects of sulfur pollution on rates of litter decomposition in a pine forest. Can J Bot 63:1436-1443.

Prescott CE (1995) Does nitrogen availability control rates of litter decomposition in forests? Plant Soil 168-169:83-88.

Priha O, Smolander A (1994) Fumigation-extraction and substrate-induced respiration derived microbial biomass C, and respiration rate in limed soil of Scots pine sapling stands. Biol Fertil Soils 17:301-308.

Quimet R, Camiré C, Furlan V (1995) Endomycorrhizal status of sugar maple in relation to tree decline and foliar, fine-roots, and soil chemistry in the Beauce region, Quebec. Can J Bot 73:1168-1175.

Raw F (1959) Earthworms population studies: a comparison of sampling methods. Nature 187:257.

Reddy GB, Reinert RA, Eason G (1991) Enzymatic changes in the rhizosphere of loblolly pine exposed to ozone and acid rain. Soil Biol Biochem 23:1115-1119.

Reich PB, Schoettle AW, Stroo HF, Troiano J, Amundson RG (1985) Effects of O_3 SO_2, and acid rain on mycorrhizal infection in northern red oak seedlings. Can J Bot 63:2049-2055.

Reich PB, Schoettle AW, Stroo HF, Amundson RG (1986) Acid rain and ozone influence mycorrhizal infection in tree seedlings. J Air Pollut Control Assoc 36: 724-726.

Reich PB, Schoettle AW, Stroo HF, Amundson RG (1988) Effects of ozone and acid rain on white pine (*Pinus strobus*) seedlings growth in five soils. III. Nutrient relations. Can J Bot 66(8):1517-1531.

Reynolds JW (1971) The effects of altitude, soil moisture, and soil acidity on earthworm (Oligochaeta: Acanthodrilidae and Lumbricidae) density, biomass, and species diversification in *Liriodendron tulipifera* L. stands in two areas of east Tennessee. Tenn Assoc Southeast Biol Bull 18:52.

Rice PM, Pye LH, Boldi R, O'Loughlin J, Tourangeau PC, Gordon CC (1979)

The effects of "low level SO$_2$" exposure of sulfur accumulation and various plant life responses of some major grassland species on the ZAPS sites. In: Colstrip MT, Preston M, Gullett TL (eds) Bioenvironmental impact of coal-fired power plant. 4th interim report. U.S. Environmental Protection Agency, Alexandria, VA.

Richards BN (1965) Mycorrhiza development of loblolly pine seedlings in relation to soil reaction and supply of nitrate. Plant Soil 22:187-199.

Rida M, Modaihsh AS (1988) Gypsum formation in sulphur treated calcareous soils. Arab Gulf J Sci Res (unpublished data). [As cited in Modaihsh et al. (1989)].

Roberts TM, Clarke TA, Ineson P, Grey TR (1980) Effects of sulphur deposition and nutrient leaching in coniferous forest soil. In: Hutchinson TC, Havas M (eds) Effects of Acidic Precipitation on Terrestrial Ecosystems. NATO Conference Series 1, Vol.4. Plenum Press, New York, pp 381-393.

Rochelle BP, Church MR, David MB (1987) Sulfur retention at intensively studied sites in the U.S. and Canada. Water Air Soil Pollut 33:73-83.

Rodale R (1948) Do chemical fertilizers kill earthworms? Org Garden 12:12-17.

Rodhe H, Granat L (1984) An evaluation of sulfate in European precipitation 1955-1982. Atmos Environ 18:2627-2639.

Rudawska M (1986) Sugar metabolism of ectomycorrhizal Scots pine seedlings as influenced by different nitrogen forms and levels. In: Gianinazzi-Pearson V, Gianinazzi S (eds) Mycorrhizae: Physiology and Genetics. (Proceedings, 1st European Symposium on Mycorrhizae, Dijon, 1985. CNRS-INRA, Dijon, pp 389-394.

Ruess L, Funke W (1992) Effects of experimental acidification on nematode populations in soil cultures. Pedobiologia 36:231-239.

Rühling Å, Tyler G (1991) Effects of simulated nitrogen deposition to the forest floor on the macrofungal flora of a beech forest. Ambio 20:261-263.

Saggar S, Bettany JR, Stewart JW (1981) Measurement of microbial sulfur in soil. Soil Biol Biochem 13:493-498.

Salonius PO (1990) Respiration rates in forest soil organic horizon materials treated with simulated acid rain. Can J For Res 20:910-913.

Satchell R (1955) Some aspects of earthworm ecology. In: Kevan DKMcE (ed) Soil Zoology. Butterworths, London, pp 180-201.

Schauermann J (1985) Zur Reaktion von Bodentieren nach Düngung von Hainsimsen-Buchenwäldern und Siebenstern-Fichtenforsten im Solling. Allg Forstzeitschr 43:1159-1160.

Schmidt J, Seiler W, Conrad R (1988) Emissions of nitrous oxide from temperate forest soils into the atmosphere. J Atmos Chem 6:95-115.

Shafer SR, Grand LF, Bruck RI, Heagle AS (1985) Formation of ecto-mycorrhizae on *Pinus taeda* seedlings exposed to simulated acid rain. Can J For Res 15:66-71.

Shafer SR (1988) Influence of ozone and simulated acidic rain on microorganisms in the rhizosphere of *Sorghum*. Environ Pollut 51:131-152.

Shafer SR (1992) Responses of microbial populations in the rhizosphere to deposition of simulated acidic rain onto foliage and/or soil. Environ Pollut 76:267-278.

Shah Z, Adams WA, Haven CDV (1990) Composition and activity of the microbial population in an acidic upland soil and effects of liming. Soil Biol Biochem 22:257-263.

Shannon JD, Sisterson DL (1992) Estimation of S and NO_x-N deposition budgets for the United States and Canada. Water Air Soil Pollut 63:211–235.

Shriner DS (1977) Effects of simulated rain acidified with sulfuric acid on host-parasite interactions. Water Air Soil Pollut 8:9–14.

Shriner DS, Henderson GS (1978) Sulfur distribution and cycling in a deciduous forest watershed. J Environ Qual 7:392–397.

Smolander A, Kurka A, Kitunen V, Mälkönen E (1994) Microbial biomass C and N in limed soil of Norway spruce stands. Soil Biol Biochem 26:503–509.

Smolander A, Mälkönen E (1994) Microbial biomass C and N, and respiratory activity in soil of repeatedly limed and N- and P-fertilized Norway spruce stands. Soil Biol Biochem 26:957–962.

Sohlenius B, Wasilewska L (1984) Influence of irrigation and fertilization on the nematode community in a Swedish pine forest soil. J Appl Ecol 21:327–342.

Sohlenius B, Bostrom S (1986) Short-term dynamics of nematode communities in arable soil — influence of nitrogen fertilization in barley crops. Pedobiologia 29: 183–191.

Speir TW, Ross DJ (1981) A comparison of the effects of air-drying and acetone dehydration on soil enzyme activities. Soil Biol Biochem 13:225–229.

Stams AJM, Booltink HWG, Lutke-Schipholt IJ, Beemsterboer B, Woittiez JRW, Van Breemen N (1991) A field study on the fate of ^{15}N-ammonium to demonstrate nitrification of atmospheric ammonium in an acid forest soil. Biogeochemistry (Dordrecht) 13:241–255.

Standen V (1984) Production and diversity of enchytraeids, earthworms and plants in fertilized hay meadow plots. J Appl Ecol 21:293–312.

Stanko KM, Fitzgerald JM (1990) Sulfur transformations in forest soils collected along an elevational gradient. Soil Biol Biochem 22:213–216.

Steiner WA (1994) The influence of air pollution on moss-dwelling animals. 2. Aquatic fauna with emphasis on Nematoda and Tardigrada. Rev Suisse Zool 101(3):699–724.

Steiner WA (1995) The influence of air pollution on moss-dwelling animals: 5. Fumigation experiments with SO_2 and exposure experiments. Rev Suisse Zool 102(1):13–40.

Stevens PA, Wannop CP (1987) Dissolved organic nitrogen and nitrate in an acid forest soil. Plant Soil 102:137-139.

Stevenson FJ (1986) The sulfur cycle. In: Cycles of Soils. Wiley, New York, pp 285–320.

Stinner DH, Stinner BR, McCartney DA (1987) Effects of simulated acidic precipitation on soil-inhabiting invertebrates in corn systems. In: Bartuska A (ed) Proceedings of National Atmospheric Conference. USDA, Washington, DC, pp. 198–206.

Straalen NM van, Geurs M, Linden JM van der (1987) Abundance, pH preference and mineral content of Oribatida and Collembola in relation to vitality of pine forests in the Netherlands. In: Perry R, Harrison RM, Bell JNB, Lester JN (eds) Acid Rain: Scientific and Technical Advances. Selper, London, pp 674–679.

Straalen NM van, Kraak MHS, Denneman CAJ (1988) Soil microarthropods as indicators of soil acidification and forest decline in the Veluwe area, the Netherlands. Pedobiologia 32:47–55.

Strayer RF, Lin C-J, Alexander M (1981) Effect of simulated acid rain on nitrification and nitrogen mineralization in forest soils. J Environ Qual 10:547–551.

Streeter J (1989) Inhibition of legume nodule formation and N₂ fixation by nitrate. CRC Crit Rev Plant Sci 7:1–23.

Stroo HF, Alexander M (1985) Effects of simulated acid rain on mycorrhizal infection of *Pinus strobus* L. Water Air Soil Pollut 25:107–114.

Stroo HF, Alexander M (1986a) Available nitrogen and nitrogen cycling in forest soils exposed to simulated acid rain. Soil Sci Soc Am J 50:110–114.

Stroo HF, Alexander M (1986b) Role of soil organic matter in the effect of acid rain on nitrogen mineralization. Soil Sci Soc Am J 50:1218–1223.

Struwe S, Kjøller A (1994) Potential for N₂O production from beech (*Fagus silvaticus*) forest soils with varying pH. Soil Biol Biochem 26:1003–1009.

Swank WT, Douglass JE (1977) Nutrient budgets of undisturbed and manipulated hardwood forest ecosystems in the mountains of North Carolina. In: Correll DL (ed) Watershed Research in North America. Smithsonian Institution Press, Washington, DC, pp 343–363.

Tabatabai MA, Bremner JM (1970) Factors affecting soil arylsulfatase activity. Soil Sci Soc Am Proc 34:427–429.

Tabatabai MA (1985) Effect of acid rain on soil. CRC Crit Rev Environ Control 15:65–110.

Tamm CO (1976) Acidic precipitation: biological effects in soil and on forest vegetation. Ambio 5:235–238.

Tamm CO, Wiklander G, Porović B (1977) Water Air Soil Pollut 8:75.

Tate KR (1984) The biological transformation of P in soil. Plant Soil 76:245–256.

Termorshuizen AJ, Ket PC (1991) Effects of ammonium and nitrate on mycorrhizal seedlings of *Pinus sylvestris*. Eur J For Path 21:404–413.

Téreault JP, Bernier B, Fortin JA (1978) Nitrogen fertilization and mycorrhizae of balsam fir seedlings in natural stands. Nat Can (Ott) 105:461–466.

Theenhaus A, Schaefer M (1995) The effects of clear-cutting and liming on the soil macrofauna of a beech forest. For Ecol Manage 77:35–51.

Theodorou C, Bowen GD (1969) Influence of pH and nitrate on mycorrhizal associations of *Pinus radiata* D. Don. Aust J Bot 17:59–67.

Thompson GW, Medve RJ (1984) Effects of aluminum and manganese on the growth of ectomycorrhizal fungi. Appl Environ Microbiol 48:556–560.

Tishler W (1955) Effect of agricultural practice on the soil fauna. In: Kevan DKMcE (ed) Soil Zoology. Butterworths, London, pp 215–230.

Titus BD, Malcolm DC (1987) The effect of fertilization on litter decomposition in clearfelled spruce stands. Plant Soil 100:297–322.

Tuttobello R, Mill PJ (1961) The pectic enzymes of *Aspergillus niger*. 1. The production of active mixtures of pectic enzymes. Biochem J 79:51–57.

Ulrich B, Pankrath J (eds) (1983) Effects of Accumulation of Air Pollutants in Forest Ecosystems. Reidel, Boston.

Valovirta I (1968) Land molluscs in relation to acidity on hyperite hills in central Finland. Ann Zool Fenn 5:245–253.

van Breemen N, Burrough PA, Velthorst EJ, van Dobben HF, de Wit T, Ridder TB, Reijnders HFR (1982) Soil acidification from atmospheric ammonium sulphate in forest canopy throughfall. Nature 299:548–550.

Van den Burg J (1976) The influence of nitrogen content of the organic matter in sandy soils without calcium on growth of coniferous species, in dependence of the phosphate and water availability. Internal report 87. Research Institute for Forestry and Landscape Planning 'De Dorschkamp'.

vanLoon GW (1984) Acid rain and soil. Can J Physiol Pharmacol 62:991–997.

vanLoon GW, Hay GW, Goh RH-T (1987) Analysis of sulfur-containing components of a soil treated with simulated acid rain. Water Air Soil Pollut 34:233–
240.

Verhoef HA, Dorel FG, Zoomer HR (1989) Effects of nitrogen deposition on
animal-mediated nitrogen mobilization in coniferous litter. Biol Fertil Soils 8:
225–259.

Vilkamaa P, Huhta V (1986) Effects of fertilization and pH on communities of
Collembola in pine forest soil. Ann Zool Fenn 23:167–174.

Vincent JM (1974) Root-nodule symbiosis with *Rhizobium*. In: Quispel A (ed) The
Biology of Nitrogen Fixation. Plenum Press, Amsterdam, pp 342–378.

Visser S, Danielson RM, Parr JF (1987) Effects of acid-forming emissions on soil
microorganisms and microbially-mediated processes. ADRP-B-02-087. Acid Deposition Research Program, Calgary, Alberta, Canada.

Visser S, Parkinson D (1989) Microbial respiration and biomass in soil of a lodgepole pine stand acidified with elemental sulfur. Can J For Res 19:955–961.

Voelker J (1959) Der chemishe Einfluss von Kalziumkarbonat auf Wachstum, Ent-
wicklung und Gehäusebau von *Achatina fulica* Bowd. (Pulmonata). Mitt Hambg
Zool Mus Inst 57:37–78.

Voigh GK (1980) Acid precipitation and soil buffering capacity. In: Drablos D,
Tollan A (eds) Ecological Impact of Acid Precipitation. Proceedings of an Inter-
national Conference, Sandefjord, Norway, March 11–14. SNSF Project, Olso,
Norway, pp 53–57.

von Lützov M, Zelles L, Scheunert I, Ottow JCG (1992) Seasonal effects of liming,
irrigation, and acid precipitation on microbial biomass N in spruce (*Picea abies*
L.) forest soil. Biol Fertil Soil 13:130–134.

Waide JB, Swank WT (1987) Patterns and trends in precipitation and stream chem-
istry at the Coweeta Hydrologic Laboratory. In: Aquatic Effects Task Group VI
Peer Review Summaries, Vol. II. North Carolina State University Atmospheric
Impacts Research Program, Raleigh, pp 421–430.

Wainwright M (1979) Microbial S-oxidation in soils exposed to heavy atmospheric
pollution. Soil Biol Biochem 11:95–98.

Wainwright M (1980) Effect of exposure to atmospheric pollution on microbial
activity in soil. Plant Soil 55:199–204.

Wainwright M, Nevell W (1984) Microbial transformations of sulphur in atmo-
spheric-polluted soils. Rev Environ Health 4:339–356.

Waldén HW (1981) Communities and diversity of land molluscs in Scandinavian
woodlands. 1. High diversity communities in taluses and boulder slopes in SW
Sweden. J Conchol 30:351–372.

Walker RF (1989) *Pisolithus tinctorius*, a Gasteromycete, associated with Jeffrey
and Sierra lodgepole pines on acid mine spoils in the Sierra Nevada. Great Basin
Nat 49:111–112.

Walker RF, McLaughlin SB (1991) Growth and root system development of white
oak and loblolly pine as affected by simulated acidic precipitation and ectomycorrhizal inoculation. For Ecol Manage 46:123–133.

Wallander H, Nylund J-E (1991) Effects of excess nitrogen on carbohydrate concentrations and mycorrhizal development of *Pinus sylvestris* L. seedlings. New Phytol 119:405–411.

Wallander H, Nylund J-E (1992) Effects of excess nitrogen and phosphorus starvation on the extramatrical mycelium of ectomycorrhizas of *Pinus sylvestris* L. New Phytol 120:495–503.

Wallwork JA (1967) Acari. In: Burges A, Raw F (eds) Soil Biology. Academic Press, New York, pp 363–395.

Wallwork JA (1970) Ecology of Soil Animals. McGraw-Hill, New York.

Wang WC, Yung YL, Lacis AA, Mo T, Hansen JE (1976) Greenhouse effects due to man-made perturbations of trace gases. Science 194:685–690.

Wäreborn I (1979) Reproduction of two species of land snails in relation to calcium salts in the foerna layer. Malacologia 18:177–180.

Wäreborn I (1982) Environments and molluscs in a non-calcareous forest area in southern Sweden. Dissertation, University of Lund.

Wäreborn I (1992) Changes in the land mollusc fauna and soil chemistry in an inland district in southern Sweden. Ecography 15:62–69.

Wästerlund I (1982) Försvinner tallens mykorrhizasvampar vid gödsling? Sven Bot Tidskr 76:411–417.

Watson AP, Van Hook RI, Jackson DR, Reichle DE (1976) Impact of a lead mining-smelting complex on the forest-floor litter arthropod fauna in the new lead belt region of southeast Missouri. ERD #881. Oak Ridge National Laboratory, Oak Ridge, TN.

Weller DE, Peterjohn WT, Goff NM, Correll DL (1986) Ion and acid budgets for a forested Atlantic coastal plain watershed and their implications for the impacts of acid deposition. In: Watershed Research Perspectives. Smithsonian Institution Press, Washington, DC, pp 392–421.

Whelpdale DM, Barrie LA (1982) Atmospheric monitoring network operations and results in Canada. Water Air Soil Pollut 18:7–23.

Wilhelmi V, Rother GM (1990) The effect of acid rain, soil temperature and humidity on C-mineralization rates in organic soil layers under spruce. Plant Soil 121: 197–202.

Will ME, Graetz DA, Roof BS (1986) Effect of acid precipitation on soil microbial activity in a Typic Quartzipsamment. J Environ Qual 15:399–402.

Williams BL (1983) Nitrogen transformations and decomposition in litter and humus from beneath closed-canopy Sitka spruce. Forestry (Oxford) 56:17–32.

Williams ML, Atkins DHF, Bower JS, Campbell GW, Irwin JG, Simpson D (1989) A preliminary assessment of the air pollution climate of the UK. Rep LR 723 (AP). Warren Spring Laboratory, Stevenage, UK.

Williams RS (1988) Effect of dilute acid on the accelerated weathering of wood. J Air Pollut Control Assoc 38:148–151.

Wisniewski J, Keitz EL (1983) Acid rain deposition patterns in the continental United States. Water Air Soil Pollut 19:327–339.

Wodzinski RS, Labeda DP, Alexander M (1978) Effects of low concentrations of bisulfite-sulfite and nitrite on microorganisms. Appl Environ Microbiol 35:718–723.

Wood M (1986) Aluminum toxicity to rhizobia. In: Megušar F, Gantar M (eds) Perspectives in Microbial Ecology. Slovene Society for Microbiology, Ljubljana, pp 659–663.

Wood M, Cooper JE (1988) Acidity, aluminium and multiplication of *Rhizobium trifolii*: effects of initial inoculum density and growth phase. Soil Biol Biochem 20:83–87.

Wookey PA (1988) Effects of dry-deposited sulphur dioxide on the decomposition of forest leaf litter. Ph.D. thesis, University of Lancaster, UK.

Wookey PA, Ineson P, Mansfield TA (1990) Effects of atmospheric sulfur dioxide on microbial activity in decomposing forest litter. Agric Ecosyst Environ 33:263–280.

Xian X, Shokohifard GI (1989) Effect of pH on chemical forms and plant availability of cadmium, zinc, and lead in polluted soils. Water Air Soil Pollut 45:265–273.

Zajonc I (1975) Variations in meadow associations of earthworms caused by the influence of nitrogenous fertilizers and liquid-manure irrigation. In: Vanek J (ed) Progress in Soil Zoology. Academia, Prague, pp 497–503.

Žel J, Gogala N (1990) Physiol Plant 79:A133.

Žel J, Blatnic A, Gogala N (1992) In vitro aluminum effects on ectomycorrhizal fungi. Water Air Soil Pollut 63:145–153.

Zelles L, Scheunert I, Kreutzer K (1987a) Effects of artificial irrigation, acid precipitation and liming on the microbial activity in soil of a spruce forest. Biol Fertil Soil 4:137–143.

Zelles L, Scheunert I, Kreutzer K (1987b) Bioactivity in limed soil of a spruce forest. Biol Fertil Soil 3:211–216.

Zelles L, Stepper K, Zsolnay A (1990) The effect of lime on microbial activity in spruce (*Picea abies* L.) forests. Biol Fertil Soil 9:78–82.

Zemba SG, Golomb D, Fay JA (1988) Wet sulfate and nitrate deposition patterns in eastern North America. Atmos Environ 22(12):2751–2761.

Manuscript received May 20, 1993; accepted May 18, 1996.

Silver Contamination in Aquatic Environments

A.R. Flegal*, I. Rivera-Duarte*, and S.A. Sañudo-Wilhelmy†

Contents

I. Introduction

Within this decade, measurements have revealed low (picomolar, p$\underline{M}$), but still relatively elevated, concentrations of dissolved (< 0.45 μm) silver in some aquatic environments. This is due to the extremely low (≤ 1 p$\underline{M}$ $\approx$ 0.1 parts per trillion) concentrations of silver in remote oceanic surface waters and a small freshwater lake compared to the relatively high concentrations (≥ 300 p$\underline{M}$) of silver in some embayments, estuaries, and fresh waters. The relatively high silver concentrations commonly correspond with elevated concentrations of silver in sediments and aquatic organisms in those water bodies. These, in turn, often have received relatively large anthropogenic inputs of industrial silver.

The two-orders-of-magnitude range in silver concentrations in aquatic habitats has raised concerns that some aquatic organisms are being exposed to toxic concentrations of silver (Flegal and Sañudo-Wilhelmy 1993; Luoma et al. 1995). While those concerns are partially substantiated by a few preliminary studies that indicate silver is extremely toxic to some invertebrates (Berthet et al. 1992; Bryan 1984; Bryan and Langston 1992; Johansson et al. 1986; Luoma and Phillips 1988; Martin et al. 1984), the validity of many silver toxicity studies is suspect (Luoma et al. 1995). This is be-

---

*Environmental Toxicology, WIGS, University of California, Santa Cruz, CA 95064, U.S.A.

†Marine Sciences Research Center, State University of New York, Stony Brook, NY 11794-5000, U.S.A.

© 1997 by Springer-Verlag New York, Inc.
Reviews of Environmental Contamination and Toxicology, Vol. 148.

cause most bioassays have not measured dissolved silver concentrations with trace-metal-clean techniques, not measured or calculated the species of dissolved silver, and not used appropriate concentrations of silver.

Consequently, this review is limited to a brief summary of the levels of silver contamination in aquatic environments. It is further abbreviated to a review of a few recent measurements of dissolved silver concentrations that have utilized trace-metal-clean techniques since these are required for accurate measurements of picomolar (pM) concentrations of elements (e.g., silver) in ambient waters (Benoit 1994; Berman and Yeats 1985; Bruland 1983; Flegal and Coale 1989; Patterson and Settle 1976; Windom et al. 1991). For silver, the peer reviewed literature is limited to the initial measurements of silver in oceanic waters of the Pacific, which were reported by Martin et al. (1983) and Murozumi (1981) in the previous decade, and a few measurements of silver in other aquatic environments, which have been reported within the present decade.

II. Background

The surprising dearth of historic information on silver contamination in the aquatic environment is presumably due to previous perceptions that it is negligible. Since it is a rare and precious metal, there have been no intentional discharges of large amounts of silver waste to aquatic environments, in contrast to some other contaminants. For example, large amounts of lead have been released into the environment for millennia as a waste product in the extraction of silver from galena by cupellation (Patterson 1971, 1978), while most industrial uses of silver (e.g., in jewelry and silverware, photography, coinage, electroplating, brazing, mirrors, dentistry, and batteries) have included extensive efforts to recover it from the effluent (Silver Institute 1994). Consequently, the largest amounts of silver released into the environment, on a global scale, appear to be from its emission as a trace constituent ($< 3 \mu g/g$) in the combustion of fossil fuels (e.g., coal, oil, and peat) and as a trace constituent of other metals (e.g., copper and zinc) during their refinement (Smith and Carson 1977).

However, there has been a pronounced increase in information on both silver contamination and toxicity in the aquatic environment within the past few years. This has resulted from the use of trace-metal-clean techniques to measure silver in natural and contaminated waters and from expanded efforts to resolve the toxicity of silver to aquatic organisms. This proliferation in information has also been catalyzed by three international conferences on the transport, fate, and effects of silver in the environment. The conferences, organized by Andren (1993, 1994, 1995), have been supported by a consortium of academic institutions, regulatory agencies, and industries. While data from those proceedings have not been cited in this report because they have not gone through a peer review process, the proceedings do contain some of the most recent and comprehensive information on silver contamination and toxicity in aquatic environments.

In addition to the numerous reports of silver toxicity to bacteria in medical and microbiology journals, there is a relatively recent review of silver toxicity to larger organisms (Luoma et al. 1995). This excellent review includes 53 references that focus on factors influencing the fate, bioavailability, and toxicity of silver in marine and estuarine environments. These references include several reports on bioassays of estuarine and marine species and life stages that are highly susceptible to silver toxicity (Calabrese et al. 1973, 1977; Dinnel et al. 1982; Eyster and Morse 1984; Lussier and Cardin 1985; Nelson et al. 1983; Sanders et al. 1990; Soyer 1963; Wilson and Freeberg 1980; Zoto and Robinson 1985), which report toxicities at silver concentrations ranging from <10 to 130 n$\underline{M}$ (<1–14 μg/L). Those reported toxic concentrations are then put in the following perspective by Luoma et al. (1995).

> "These are low concentrations for the toxicity of any trace element. Nevertheless, widespread concentrations of Ag in the μg L^{-1} range are rarely reported even in polluted natural water. The highest concentrations of dissolved Ag reported (using reliable chemical techniques) in open waters of San Francisco Bay are 0.025 μg L^{-1} (Smith and Flegal 1993). Thus, a disparity exists between experimental toxicity and natural dissolved concentrations of Ag. This is balanced by the likelihood that toxicity tests, because of their simplistic designs, are less sensitive than natural systems. For example, the tests do not employ the most sensitive species from nature (and sensitive life processes are not fully known). Experimental toxicity tests rarely account for multiple pathways of exposure, nor can most designs consider the complex, often secondary interactions that influence toxicity in ecosystems (Luoma, in press). The toxicological uncertainties imposed by chemical disparities and biological insensitivities suggest that no conclusion can yet be drawn about Ag toxicity in polluted waters."

In summary, orders-of-magnitude variations in total dissolved silver concentrations have heightened concerns about its potential toxicity to aquatic organisms. Those concerns have been exacerbated by the absence of recognized natural levels of silver in different aquatic environments. Therefore, this review includes both a summary of the range of silver concentrations reported for different water bodies and estimates of their natural dissolved silver concentrations. Together, these provide a perspective on the levels of silver contamination found in different aquatic environments.

III. Natural Levels of Silver in Oceanic Waters

The natural concentrations of silver in oceanic waters are illustrated in Fig. 1, which is based on vertical profiles of silver concentration gradients in remote regions of the Northeast Pacific (Martin et al. 1983) and the Northeast Atlantic (Flegal et al. 1995). Comparisons of the Pacific profiles of total dissolved (<0.45-μm-diameter) silver and the Atlantic profiles of total (total dissolved and particulate) silver are based on measurements indicating that essentially all ($>80\%$) silver in oceanic waters is <0.45 μm (Martin et al. 1983). The systematic variations in those profiles are characteristic of the geochemical cycling of nutrient-type elements from the North Atlantic

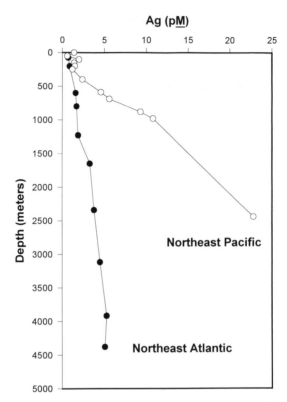

Fig. 1. Vertical profile concentrations of total (dissolved and particulate) silver (Ag) (pM) in the Northeast Atlantic (Flegal et al. 1995) and total dissolved ($<$0.45-μm-diam.) silver in the Northeast Pacific (Martin et al. 1983). The profiles are comparable based on measurements (Martin et al. 1983) indicating that dissolved silver is the predominant ($\geq$80%) phase in oceanic waters.

to the North Pacific, which is described in detail elsewhere (Broecker and Peng 1982).

The nutrient-type cycling of silver in oceanic waters is evidenced by its positive covariance with silicate. This is illustrated by the following, highly significant ($p < 0.001$), simple linear regressions of silver [Ag] and silicate [H$_4$SiO$_4$] concentrations in the Atlantic (Flegal et al. 1995) and in the Pacific (Martin et al. 1983).

Equation 1: Ag in the Atlantic Ocean:

$$\text{Ag (p\underline{M})} = 0.685 + 0.107 \times \text{H}_4\text{SiO}_4 \ (\mu\underline{M}) \ (R = .916)$$

Equation 2: Ag in the Pacific Ocean:

$$\text{Ag (p\underline{M})} = -0.691 + 0.111 \times \text{H}_4\text{SiO}_4 \ (\mu\underline{M}) \ (R = .897)$$

Several factors contribute to differences in the intercepts of the two regressions. Some of the principal factors are these: the number of analyses in both data sets is small; there is analytical imprecision in measurements of silver concentrations as they approach zero (<1 p$\underline{M}$) in oceanic surface waters; and there are physical, biological, chemical, and geological differences within and between oceanic water masses that may alter the ratio of silver and silicate. Therefore, additional measurements are required to distinguish between analytical and natural variations in [Ag]:[H_4SiO_4] ratios in oceanic waters.

The utility of the natural covariance of silver and silicate in oceanic waters as a measure of contamination is two-fold. First, the systematic increase of silver and silicate concentrations in the oceans indicates that the distribution of silver in the open ocean is governed by natural biogeochemical processes that have not been markedly perturbed by anthropogenic processes. Second, anthropogenic perturbations of the silver cycle in coastal (neritic) waters may be quantified by deviations from those oceanic [Ag]:[H_4SiO_4] correlations. Consequently, the natural concentrations of silver in oceanic surface waters are considered to range from ≤ 0.7 p$\underline{M}$ in the Atlantic to ≤ 2 p$\underline{M}$ in the Pacific, and natural concentrations of silver in subsurface waters are considered to be ≤ 10 p$\underline{M}$ in the Atlantic and ≤ 23 p$\underline{M}$ in the Pacific.

Based on the preceding characterizations, silver concentrations in remote surface waters of the North Atlantic may be relatively contaminated. Surface water concentrations in the two vertical profiles in the North Atlantic ranged from 0.25 to 0.70 p$\underline{M}$, while surface water concentrations in the two vertical profiles in the South Atlantic were ≤ 0.24 p$\underline{M}$ (Flegal et al. 1995). The apparent hemispheric disparity in surface water concentrations suggests that ambient silver concentrations in some North Atlantic surface waters may be elevated two- or threefold above their natural concentrations by the atmospheric deposition of industrial silver aerosols advected by prevailing westerly and easterly winds. This hypothetical level of silver contamination is based on parallels with the documented level of lead contamination in the North Atlantic (Veron et al. 1994).

The apparent enrichment of silver in the North Atlantic surficial waters may also be attributed to aeolian inputs of natural silver in Saharan dust advected to the North Atlantic (Flegal et al. 1995). This hypothetical natural enrichment is consistent with complementary measurements of aluminum in those waters (Measures 1995). Since silver concentrations in oceanic surface waters of the North Atlantic are still less than natural silver concentrations in oceanic surface waters of the North Pacific, it may be conservatively concluded that natural concentrations of silver in oceanic surface waters are ≤ 2 p$\underline{M}$.

IV. Silver Contamination in Neritic Waters

There are substantially elevated ($\approx$ 10-fold) concentrations of silver (up to 40 p$\underline{M}$) in some coastal or neritic surface waters, relative to adjacent oceanic surface waters, from point-source discharges of industrial silver to coastal waters (Bloom and Crecelius 1984, 1987; Sañudo-Wilhelmy and Flegal 1992). This perturbation is evidenced by excesses of silver relative to silicate within the Southern California Bight, using the preceding regression for their covariance (Eq. 2) in the adjacent North Pacific. The excesses occur near ocean outfalls off Tijuana, Mexico, and San Diego, California, U.S.A., which discharge relatively large amounts of silver to the Southern California Bight (≈ 25 tonnes/yr).

The initial measurements of elevated silver concentrations in some neritic waters are consistent with other reports of silver contamination in the coastal environment. These include reports of elevated concentrations of silver in marine sediments (e.g., Bloom and Crecelius 1987; Bothner et al. 1994; Bryan 1984, 1985; Bryan et al. 1985; Hornberger et al., in manuscript; Luoma and Bryan 1978, 1981; Luoma et al. 1990, 1995; O'Connor 1992; Rutherford and Church 1975) and marine invertebrates (e.g., Alexander and Young 1976; Berthet et al. 1992; Bryan and Hummerstone 1977; Bryan and Langston 1992; Calabrese et al. 1984; Cherry et al. 1983; Flegal 1980; Goldberg et al. 1983; Luoma and Bryan 1982; Luoma et al. 1995; Martin et al. 1984, 1988; O'Connor 1992; Thomson et al. 1984) collected near point-source discharges. In total, those parallel measurements of elevated silver concentrations near wastewater outfalls attest to the industrial origins of silver excesses in neritic waters.

V. Silver Contamination in Estuarine Waters

A preliminary diagnostic measure of silver contamination in estuarine waters may be derived from a plot of dissolved silver concentrations relative to salinity (Fig. 2). Excesses in silver relative to salinity within an estuarine gradient, which are characteristic of a nonconservative input of silver within the estuary, have been observed in California (Smith and Flegal 1993), Texas (Benoit et al. 1994), and New York (Sañudo-Wilhelmy, manuscript in preparation) estuaries. While those excesses may result from inputs (atmospheric, point-source, and non-point-source) of anthropogenic silver (Davis et al. 1992), they may also be due to the natural desorption of silver from estuarine sediments (Davis 1977). The latter process is indicated by the covariance of truly dissolved (< 10 kDa) silver and aluminum in the San Francisco Bay estuary (Sañudo-Wilhelmy et al. 1997), which is attributed to the diagenetic remobilization of silver from contaminated sediments.

The relationship between elevated concentrations of silver in surface waters and benthic sediments was first documented with analyses of the distribution of silver in two semienclosed embayments (Flegal and Sañudo-Wilhelmy 1993). Comparable levels of contamination were observed in surface waters of San Diego Bay and South San Francisco Bay, where the dissolved silver concentrations ranged up to 300 p$\underline{M}$. The elevated concentrations corresponded with elevated silver concentrations in benthic sediments in both embayments (Luoma and Phillips 1988). The sediments also

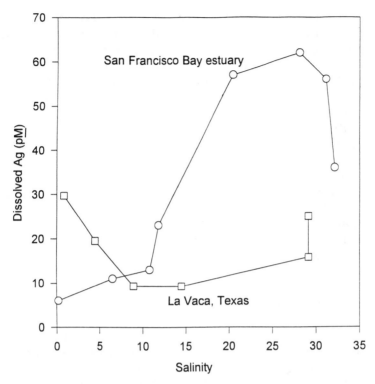

Fig. 2. Nonconservative distributions in total dissolved (<0.45-μm-diam.) silver with salinity in two estuaries: San Francisco Bay, California (from Smith and Flegal 1993) and La Vaca, Texas (from Benoit et al. 1994). Silver concentrations above the mixing line defined by the freshwater and saltwater end members evidence an internal source of silver (e.g., San Francisco Bay), while concentrations below that line evidence an internal sink of silver (e.g., La Vaca) within an estuary.

appeared to be the primary source of silver contamination within San Diego Bay, because all point-source discharges to that bay were terminated three decades ago. This suggested that benthic fluxes of silver from previously contaminated sediments could be a persistent source of contamination in aquatic environments with limited hydraulic flushing (Flegal and Sañudo-Wilhelmy 1993).

The potential importance of inputs of silver from contaminated sediments to overlying waters within embayments was illustrated by mass-balance calculations of silver fluxes in the San Francisco Bay estuary (Smith and Flegal 1993). These data indicated that the benthic flux of silver from diagenetic remobilization from contaminated sediments in South San Francisco Bay was ≈ 1.7 μg/m^2 d (235 kg/yr) and that the benthic flux of dissolved silver throughout the entire estuary was ≈ 1.2 μg/m^2 d (540 kg/

yr). On a systemwide scale, the estimated benthic flux was an order of magnitude greater than the fluvial flux of silver to the estuary (12 kg/yr).

Those initial estimates were substantiated by an independent estimate of benthic fluxes in the estuary (Rivera-Duarte and Flegal 1996). The upper limits of the integrated net benthic flux (from both physical diffusion and biological irrigation of the sediment pore waters) of silver to South Bay were estimated to be 3–31 kg/yr. That highly qualified estimate indicates the net diffusive flux of silver could be up to 2.5 times the entire riverine input of dissolved silver to San Francisco Bay. The estimated diffusive benthic fluxes of silver were greatest at sites affected by wastewater outfalls with relatively elevated concentrations of silver in the effluent, where sediment concentrations are 15-fold greater than background concentrations in the estuary (Luoma et al. 1985). This validated the proposal that contaminated sediments could be a relatively important source to overlying waters in semienclosed embayments with limited hydraulic flushing.

Comparable concentrations of dissolved silver have since been reported for waters in six Texas estuaries (Benoit et al. 1994). The average dissolved (<0.4 μm) silver concentration in the Texas estuaries was commonly less than the relatively high detection limit of 1.0 ng/kg (9 p$\underline{M}$) during the fall. However, the average ($\bar{x} \pm$ SD) concentrations were markedly higher in the Sabine (65 $\pm$ 10 p$\underline{M}$), Galveston (72 $\pm$ 9 p$\underline{M}$), San Antonio (37 $\pm$ 6 p$\underline{M}$), and Corpus Christi (39 $\pm$ 8 p$\underline{M}$) estuaries during the summer. These analyses substantiate the proposal that silver concentrations in some estuarine waters may be one or two orders of magnitude above natural oceanic concentrations.

Additionally, the seasonal increases of silver concentrations in the Texas estuaries (Benoit et al. 1994) were consistent with seasonal increases of silver concentrations in San Francisco Bay (Smith and Flegal 1993). The highest concentrations (240 p$\underline{M}$) there also occurred in the summer months. This was consistent with the proposal that the greatest elevations of silver concentrations in estuarine waters occur during periods of relatively low freshwater discharge (hydraulic flushing) and relatively high benthic activity, which maximize benthic inputs from diagenetic remobilization (Flegal and Sañudo-Wilhelmy 1993).

In summary, silver concentrations of several estuaries appear to be elevated by inputs from several anthropogenic sources. These include freshwater discharges, surface runoff, point-source discharges, and benthic fluxes from contaminated sediments. (Atmospheric inputs of industrial silver from fossil fuel combustion, smelting, and cloud seeding, which are relatively large on a global scale, are not considered to contribute to localized increases in silver concentrations in relatively small water bodies.) Although there is no currently recognized natural concentration of silver in estuaries, these comparisons indicate that silver concentrations in some estuaries are two orders of magnitude above natural concentrations. These are tentatively estimated to be ≤ 5 p$\underline{M}$.

However, natural concentrations of silver in estuaries are expected to vary markedly with other parameters that influence its speciation. This is illustrated in Fig. 3, which shows pronounced differences in the percentage of total dissolved silver (<0.45 μm) in colloidal fractions (>10 kDa) in the Hudson River and San Francisco Bay estuaries. The differences are attributed to higher concentrations of colloidal material in the Hudson, which increase its complexing capacity for silver.

VI. Silver Contamination in Fresh Waters

Benoit (1994) conducted the first extensive study, using trace-metal-clean techniques, of dissolved silver in freshwater systems. His analyses of dissolved (<0.45 μm) silver in the Quinnipiac River (Connecticut, USA) ranged from ≤ 9 p$\underline{M}$ to $2.8 \mu\underline{M}$ (2,800,000 p$\underline{M}$). The total (unfiltered) silver concentrations in that river ranged from ≤ 9 p$\underline{M}$ to 4.6 $\mu\underline{M}$ (4,600,000 p$\underline{M}$).

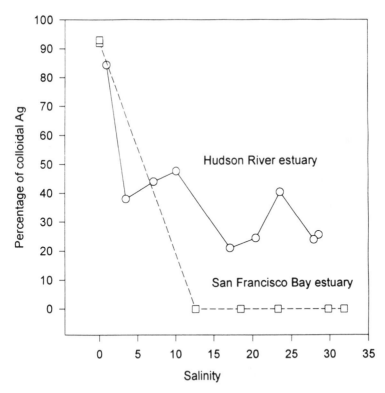

Fig. 3. Variations of total dissolved (<0.45-μm-diam.) silver in colloidal phases (>10 kDa) with salinity in the Hudson River (Sañudo-Wilhelmy, unpublished data) and San Francisco Bay estuaries (Sañudo-Wilhelmy et al. 1997). Both sets of data were acquired with the same sampling system and analytical techniques.

These extremely wide ranges in both dissolved and total silver concentrations were attributed to differences in silver concentrations between relatively pristine and highly contaminated fresh waters.

Representative total dissolved silver concentrations in those and other riverine systems are plotted in Fig. 4. The lower range in those concentrations suggests that natural concentrations of silver in freshwater systems are ≤ 10 pM. Based on that proposed baseline concentration, silver concentrations in some highly contaminated rivers appear to be orders of magnitude greater than natural concentrations.

The proposed natural concentration of silver in freshwater systems (≤ 10 pM) is supported by preliminary measurements of total dissolved silver concentrations in a small, freshwater impoundment, Davis Creek, in California (Sañudo-Wilhelmy and Flegal, unpublished data). The vertical profile of silver concentrations (< 1–3 pM) in that shallow reservoir brackets the concentrations of silver (≈ 1 pM) at corresponding depths in the Atlantic (Fig. 5). This suggests that natural silver concentrations in freshwater systems are comparable to those in oceanic surface waters, as is true for lead (Erel and Patterson 1994; Flegal and Coale 1989; Flegal et al. 1989).

As in estuaries, natural differences in dissolved silver concentrations between rivers may be partially due to differences in their complexation

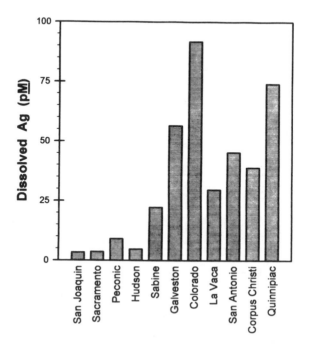

Fig. 4. Representative total dissolved (< 0.45-μm-diam.) silver concentrations (pM) in 11 U.S. rivers (from Benoit 1994; 1995; Benoit et al. 1994; Sañudo-Wilhelmy et al., unpublished data; Smith and Flegal 1993).

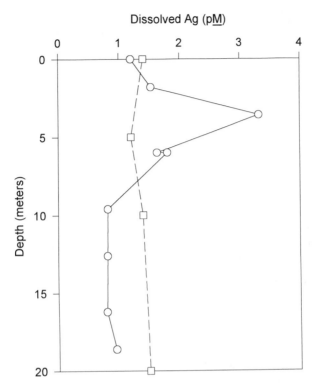

Fig. 5. Total dissolved (<0.45-μm-diam.) silver concentrations (p$\underline{M}$) in surface waters of the Northeast Atlantic (*squares*) and a small freshwater impoundment, Davis Creek (*circles*), in California. The ocean data are from Flegal et al. (1995); the freshwater data are from unpublished measurements using the same techniques (Sañudo-Wilhelmy and Flegal, unpublished data).

capacity (Benoit 1994; Cowan et al. 1985; Davis 1977; Engel et al. 1981; Luoma et al. 1995; Sañudo-Wilhelmy et al. 1997); this was indicated by the relatively high percentage ($>70\%$) of dissolved (<0.45 μm) silver in colloidal phases (10 kDa to 0.2 μm) of rivers within the United States (Sañudo-Wilhelmy et al. 1996; Sañudo-Wilhelmy, unpublished data) and Japan (Tanizaki et al. 1992). That colloidal partitioning is shown in Fig. 6.

VII. Silver Contamination in Sediment Pore Waters

While there are very few reports of dissolved (<0.45 μm) silver concentrations in sediment pore waters (Lyons and Fitzgerald 1983; Rivera-Duarte and Flegal 1996), these do provide unique insights on the biogeochemical cycling of silver in aquatic environments. The reports indicate (i) the importance of chloro-complexation on the solubilization of silver in natural sediment pore waters and (ii) the association of silver with anthropogenic

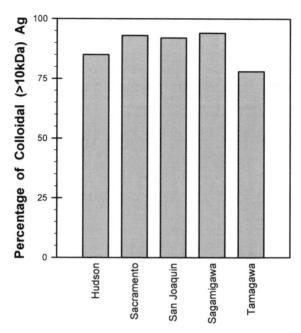

Fig. 6. Percentage of total dissolved (<0.45-μm-diam.) silver in colloidal phases (>10 kDa) in the Hudson, Sacramento, and San Joaquin rivers in the United States (Sañudo-Wilhelmy, unpublished data) and in the Sagamigawa and Tamagawa rivers in Japan (Tanizaki et al. 1992).

sources. This is depicted by the vertical concentration profiles of dissolved silver in Long Island Sound, San Francisco Bay, Tomales Bay, and Drakes Estero (Fig. 7).

Although relatively high concentrations (4500 p$\underline{M}$) were found in pore waters near a large wastewater outfall (Mayfield Slough) in San Francisco Bay, the highest dissolved silver concentrations (up to 9000 p$\underline{M}$) of silver were observed in pore waters from surficial sediment in relatively pristine sites (Mystic River Estuary in Long Island Sound and Tomales Bay). In contrast, pore water concentrations in Brandford tidal flat (Long Island Sound) were <2000 p$\underline{M}$, and most of the sites sampled within San Francisco Bay and Drakes Estero had concentrations of <300 p$\underline{M}$. The highest concentrations are tentatively attributed to chloro-complexation of silver in those pore waters and the reduction of iron and manganese oxides rather than to anthropogenic contamination (Lyons and Fitzgerald 1983; Rivera-Duarte and Flegal 1996). (This attribution is based on thermodynamic calculations on the speciation of silver in aquatic environments, which are briefly summarized in the following section.) Consequently, it is relatively difficult to estimate a natural concentration of silver in sediment pore waters or to quantify levels of silver contamination in those waters.

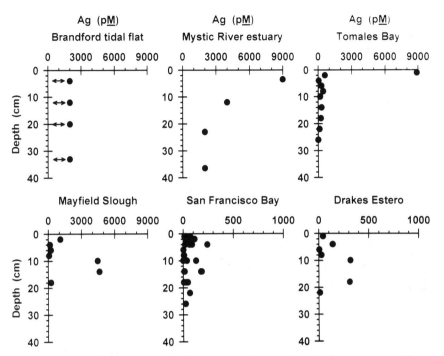

Fig. 7. Vertical profile concentrations (p$\underline{M}$) of total dissolved (< 0.45-μm-diam.) silver in sediment pore waters from the Brandford tidal flat and Mystic River estuary in Long Island Sound on the east coast of the United States (Lyons and Fitzgerald 1983) and from San Francisco Bay, Drakes Estero, and Tomales Bay on the west coast of the United States (Rivera-Duarte and Flegal 1996). Since silver concentrations in Mayfield Slough, which is a highly contaminated site in the southern end of San Francisco Bay, are markedly higher than those at other sites within the estuary, those concentrations are plotted separately.

VIII. Speciation of Dissolved Silver in Aquatic Environments

The chemical speciation of silver and the processes affecting its adsorption to particles strongly influence its bioavailability in natural waters. Thermodynamic calculations (see following) indicate that, in the absence of sulfides (H_2S and HS^-), silver is mainly present as the monovalent silver ion (Ag^+) in fresh water (Jenne et al. 1978; Whitlow and Rice 1985), but silver is predominantly a neutral silver bisulfide complex ($AgHS°$) in fresh water with sulfide levels as low as 1×10^{-5} mg/L. In contrast, those thermodynamic calculations indicate that the neutral silver–chloro complex ($AgCl°$) is the principal species in saline water (Cowan et al. 1985; Jenne et al. 1978). This theoretically reduces the relative adsorption of silver to particles in estuarine and marine environments (Davis 1977; Luoma et al. 1995), which is consistent with the report that silver desorption from inorganic particles increases with salinity in an estuarine system (Sanders and Abbe

1987; Sañudo-Wilhelmy et al. 1996). The chloro-complexation also influences the bioaccumulation of silver in those environments because of the relatively high bioavailability of $AgCl^\circ$ (Cowan et al. 1985; Engel et al. 1981; Luoma et al. 1995).

The preceding summary is based on a small number of detailed thermodynamic studies of the chemical speciation of silver in natural aquatic environments, which were pioneered by Jenne et al. (1978). They modeled (WATEQ2) the equilibrium distribution of silver among numerous inorganic solute complexes and solid phases and the changes expected during a transit through an estuarine system. Their model incorporated data from the San Francisco Bay estuary, from the fresh (salinity ≤ 0.13) to the marine (salinity $= 33.5$) water end-members (Figs. 8 and 9).

As a first approximation, they assumed that organic complexation of silver is not important in natural aquatic environments. Their assumption was tentatively corroborated for saline waters by Cowan et al. (1985), whose calculations with MINTEQ indicated that organically complexed silver represents $< 1\%$ of the total dissolved silver in marine waters, and by Miller and Bruland (1995), whose measurements indicated the absence of measurable organic complexation of silver in samples from both the Weddell Sea and South San Francisco Bay. However, subsequent studies on size-speciation of trace elements within San Francisco Bay indicate the association of silver in fresh water with colloidal material (10 kDa to 0.2 μm), which could be either organic or clay-size minerals (Sañudo-Wilhelmy et al. 1997; Tanizaki et al. 1992), as illustrated in Fig. 6.

The effect of chloride complexation of silver in marine waters is depicted in Fig. 8a (Jenne et al. 1978). As the salinity increases from fresh water (0.13) to marine waters (33.5), the activity (A; i.e., reactivity) of uncomplexed silver (A_{Ag^+}) decreases sharply (from 4×10^{-11} to 3×10^{-15}) while the activities of all the chloride complexes, except $AgCl^\circ$, increase. For marine waters, $A_{AgCl_2^-} > A_{AgCl_3^{2-}} > A_{AgCl_4^{3-}} > A_{AgCl^\circ}$. The preponderance of silver chloride complexes was also indicated by independent calculations (MINEQL) for pore waters in suboxic coastal sediments (Lyons and Fitzgerald 1983).

The thermodynamic models indicate that bromide and iodide complexes account for only a minor portion of the dissolved silver at the concentrations of bromide, iodide, and chloride in marine and fresh waters (Fig. 8b). Even when bromide (and iodide) concentrations are considered 10-fold larger (Fig. 8c), the most abundant bromide complex ($AgBr_2^-$) in the marine end member is only 1.7% of the activity of the most abundant chloride complex ($AgCl_2^-$), and it is equivalent to the activity of the least abundant chloride complex ($AgCl^\circ$). Iodide (Fig. 8c) and fluoride (8d) complexes are even less important in those environments.

The complexation with sulfide seems to be the most important in both fresh water and marine waters (Cowan et al. 1985; Jenne et al. 1978). Assuming concentrations of 1×10^{-5} mg/L for H_2S in the San Francisco

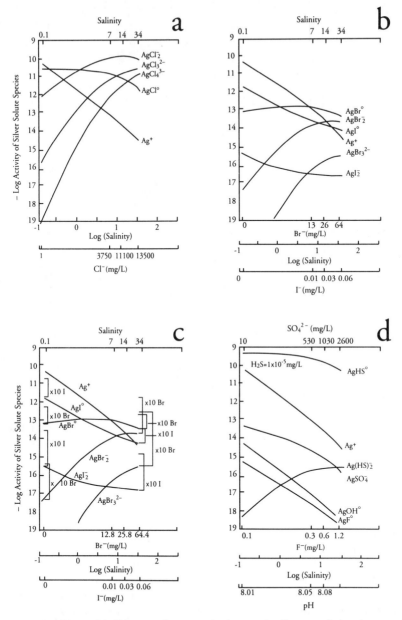

Fig. 8. The effect of halide complexes on the inorganic silver speciation in aquatic systems, with salinities ranging from fresh water (0.1) to sea water (34) (based on analyses of Jenne et al. 1978). See the text for discussion of variations in Cl^{-1}(a), Br^{-1} (b,c), I^{-1}(b,c), and F^{-1} (d).

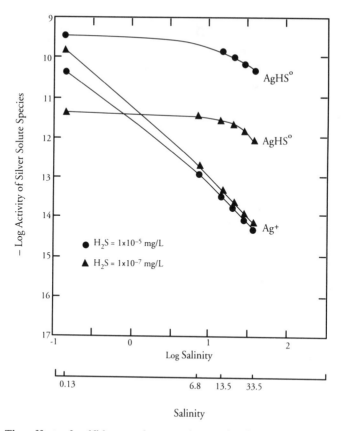

Fig. 9. The effect of sulfide complexes on inorganic silver speciation in aquatic systems, with salinities ranging from fresh water (0.1) to sea water (34) (based on analyses of Jenne et al. 1978).

Bay estuary, Jenne et al. (1978) calculated that the silver bisulfide ion pair (AgHS°) is the dominant species in the transition from the Sacramento River to the Pacific Ocean (Fig. 8d). Cowan et al. (1985) made the same conclusion for marine waters with S^{2-} concentrations above 1×10^{-5} mg/L. The silver bisulfide ion pair was also estimated to be the most important species in pore water from anoxic coastal sediment pore waters (Lyons and Fitzgerald 1983). A decrease in H_2S concentration from 1×10^{-5} to 1×10^{-7} mg/L produces a decrease of $A_{AgHS°}$ by 50-fold for marine and 100-fold for freshwater end members (Fig. 9). However, this decrease is not justified by recent measurements that indicate a concentration range of H_2S from $<3 \times 10^{-5}$ to 5×10^{-3} mg/L in waters of San Francisco Bay (Kuwabara and Luther 1993).

The speciation of dissolved silver in natural waters has been calculated with numerous thermodynamic models, as previously discussed. However, those models need to be substantiated with direct measurements. Additionally, more complex models including organic speciation have to be developed.

IX. Partition Coefficient of Silver in Natural Aquatic Environments

As previously indicated, there have been only a few measurements of total dissolved (< 0.45 μm) silver concentrations in any natural waters. Since total dissolved concentrations are operationally defined by a filter size, they include silver in small particulates, colloids, complexes, ion pairs, and free species. Consequently, analyses of total dissolved silver concentration measurements are subject both to misinterpretation and overinterpretation.

With that qualification, the ratio of total dissolved silver to particulate silver in aqueous systems provides some useful diagnostic information (Benoit 1995; Smith and Flegal 1993). The distribution is termed the partition coefficient (K_d), which is defined by the following equation:

$$K_d = \frac{[\mathrm{Ag}]_{\text{particulate}}}{[\mathrm{Ag}]_{\text{dissolved}}}$$

As the ratio often varies by orders of magnitude, it is commonly reported as a logarithm (log K_d). This is illustrated by the initial data for San Francisco Bay, where the partitioning of silver (log K_d) ranged from 4.3 to 6.1 within the water column (Smith and Flegal 1993) and from 2.1 to 5.8 within sediment pore waters (Rivera-Duarte and Flegal 1996). The latter contrasted with the lower ratios of silver (log K_d) in sediment pore waters of adjacent, relatively pristine, estuaries (Tomales Bay and Drakes Estero), which ranged from 1.2 to 4.2 (Rivera-Duarte and Flegal 1996). The differences are partially attributed to the larger grain size of the sediments and complexation of silver by chloride in these pristine embayments (Rivera-Duarte and Flegal 1996). This is consistent with previously discussed variations in pore water concentrations.

The variations in log K_d and the corresponding concentration gradients of silver in sediment pore waters indicate that contaminated sediments are a source of silver to overlying waters. The gradients attest to the remobilization of silver from particulate to dissolved phases during early sediment diagenesis. Dissolved silver within sediment pore waters may then be transported to the water column by both diffusive and advective processes. This movement would be accelerated by seasonal increases in benthic activity, which often correspond with summer periods of reduced hydraulic flushing in many aquatic environments.

The partitioning of silver is influenced by numerous natural and anthropogenic factors (Benoit 1995; Sañudo-Wilhelmy et al. 1997). By definition, the K_d of silver is inversely correlated to the total suspended sediment (TSS) load. It has also been proposed that relatively large amounts of colloids substantially lower the K_d of silver in estuarine waters, due to the particle concentration effect (Benoit 1995). While preliminary measurements (Sañudo-Wilhelmy et al. 1997) indicate that most ($\approx 84\%$) of the total dissolved silver in fresh waters discharging into the estuary is in a colloidal phase (10 kDa to 0.2 μm), there is essentially none ($<1\%$) of the total dissolved silver in a colloidal phase within San Francisco Bay (see Fig. 6). Consequently, colloids may contribute to the particle concentration effect on the K_d of silver in some fresh waters or other estuarine systems, but they do not appear to contribute to that effect in San Francisco Bay or marine waters.

X. Scavenging and Remobilization

The partitioning of silver is primarily controlled by its speciation (Cowan et al. 1985; Jenne and Luoma 1977). Davis (1977) reported a striking effect of chloride on the adsorption of silver onto freshly made iron oxides (goethite). He observed that 80%–90% of the dissolved silver was adsorbed in 15 hr in the presence of traces of chloride (10^{-3} $\underline{M}$, but $<5\%$ was adsorbed in 15 hr when the salinity was increased to ≈ 5 $(9.4 \times 10^{-2}$ $\underline{M}$ Cl$^-$). These results are concordant with those of Sanders and Abbe (1987), who observed an increase in the desorption of silver from inorganic particles as salinity increased in an estuarine system, and with measurements of high pore water concentrations of dissolved (<0.45 μm) silver in relatively pristine coastal sediments (Lyons and Fitzgerald 1983; Rivera-Duarte and Flegal 1996). In contrast, Luoma et al. (1995) reported that natural oxidized estuarine sediment adsorbed $>99\%$ of the silver from solution in 24 hr at salinity of 20.

Luoma et al. (1995) also found that the rate of adsorption was reduced when sediments were stripped of their natural reactive coatings (e.g., complex aggregates of iron oxides, manganese oxides, and organic material) and that 55%–85% of the adsorbed silver was extractable with a weak acid leach (0.5 $\underline{N}$ HCl), which solubilizes amorphous iron and manganese oxides. This is consistent with other reports of the affinity of silver to particulate manganese in fresh water (Anderson et al. 1973; Borovec 1993; Chao and Anderson 1974), and estuarine (Rivera-Duarte and Flegal 1996) sediments. Therefore, the scavenging and remobilization of silver from sediments appears to be strongly correlated with the formation and dissolution of ferromanganese (oxy)hydroxides in oxic and anoxic environments.

These and other factors influencing the bioavailability of silver in aquatic (marine and estuarine) environments are discussed by Luoma et al. (1995). As previously indicated, that review primarily addresses analyses of particulate silver concentrations in sediments and organisms, because there are relatively few measurements of silver in dissolved phases. Consequently, that review corroborates the hypothesis of this review, which is that future studies of silver bioavailability and toxicity need to incorporate trace-metal-clean techniques measuring dissolved silver concentrations and species.

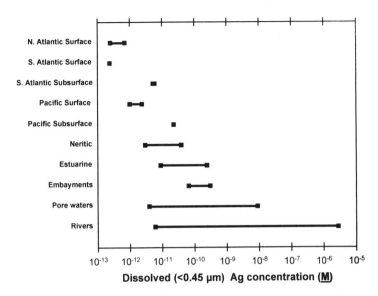

Fig. 10. Ranges of total dissolved (< 0.45-μm-diam.) silver concentrations ($\underline{M}$) in aquatic environments (based on numerous references cited in the text).

Summary

Recent measurements have revealed a surprising range in dissolved silver concentrations in aquatic environments (Fig. 10). Concentrations in some neritic, estuarine, and fresh waters are one to seven orders of magnitude above natural levels in oceanic surface waters. These levels of contamination are relatively higher than those of any other element in aquatic environments.

While the bioavailability and toxicity of silver in those contaminated aquatic environments may be relatively high, they depend on its speciation. Speciation has, to date, only been calculated with simple thermodynamic models, and there have only been a few measurements of dissolved silver species in natural waters. Furthermore, most bioassays of silver toxicity with aquatic organisms have not utilized trace-metal-clean techniques to accurately measure the concentration of dissolved silver, much less its speciation. Therefore, the potentially adverse effects of anthropogenically elevated silver concentrations in some contaminated aquatic systems are still speculative.

References

Alexander GV, Young DR (1976) Trace metals in southern California mussels. Mar Pollut Bull 7:7–9.
Anderson BJ, Jenne EA, Chao TT (1973) The sorption of silver by poorly crystallized manganese oxides. Geochim Cosmochim Acta 37:611–622.

Andren AW (ed) (1993) Proceedings of the 1st International Conference on the Transport, Fate and Effects of Silver in the Environment. University of Wisconsin Sea Grant Institute, Madison.

Andren AW (ed) (1994) Proceedings of the 2nd International Conference on the Transport, Fate and Effects of Silver in the Environment. University of Wisconsin Sea Grant Institute, Madison.

Andren AW (ed) (1995) Proceedings of the 3rd International Conference on the Transport, Fate and Effects of Silver in the Environment. University of Wisconsin Sea Grant Institute, Madison.

Benoit G (1994) Clean technique measurement of Pb, Ag, and Cd in freshwater: a redefinition of metal pollution. Environ Sci Technol 28:1987–1991.

Benoit G (1995) Evidence of the particle concentration effect for lead and other metals in fresh waters based on ultraclean technique analyses. Geochim Cosmochim Acta 59:2677–2687.

Benoit G, Oktay-Marshall SD, Cantu A II, Hood EM, Coleman CH, Corapcioglu MO, Santschi PH (1994) Partitioning of Cu, Pb, Ag, Fe, Al, and Mn between filter-retained particles, colloids, and solution in six Texas estuaries. Mar Chem 45:307–336.

Berman SS, Yeats PA (1985) Sampling of seawater for trace metals. CRC Crit Rev Anal Chem 16:1–14.

Berthet B, Amiard-Triquet JC, Martoja M, Jeantet A (1992) Bioaccumulation, toxicity and physico-chemical speciation of silver in bivalve mollusks: ecotoxicological and health consequences. Sci Total Environ 125:97–122.

Bloom NS, Crecelius EA (1984) Determination of silver in sea water by coprecipitation with cobalt pyrrolidinidithiocarbamate and Zeeman graphite-furnace atomic absorption spectrometry. Anal Chim Acta 156:139–145.

Bloom NS, Crecelius EA (1987) Distribution of silver, mercury, lead, copper, and cadmium in central Puget Sound sediments. Mar Chem 21:377–390.

Borovec Z (1993) Partitioning of silver, beryllium and molybdenum among chemical fractions in the sediments from the Labe (Elbe) River in Central Bohemia, Czech Rep. Geojournal 29(4):359–364.

Bothner MH, Takada H, Knight IT, Hill RT, Butman B, Farrington JW, Colwell RR, Grassle JF (1994) Sewage contamination in sediments beneath a deep-ocean dump site off New York. Mar Environ Res 38:43–59.

Broecker WS, Peng T-H (1982) Tracers in the Sea. Eldigio, Columbia University, New York.

Bruland KW (1983) Trace elements in sea-water. In: Riley JP, Chester R (eds) Chemical Oceanography, Vol. 8. Academic Press, New York, pp 157–220.

Bryan GW, Hummerstone LG (1977) Indicators of heavy-metal contamination in the Looe estuary (Cornwall) with particular regard to silver and lead. J Mar Biol Assoc UK 57:75–92.

Bryan GW (1984) Pollution due to heavy metals and their compounds. Mar Ecol 5:1290–1431.

Bryan GW (1985) Bioavailability and effects of heavy metals in marine deposits. In: Ketchum BH, Capuzzo JM, Burt WV, Duedall IW, Park PK, Kester DR (eds) Wastes in the ocean, Vol. 6. Nearshore Waste Disposal. Wiley & Sons, New York, pp 42–79.

Bryan GW, Langston WJ, Hummerstone LG, Burt GR (1985) A guide to the assessment of heavy-metal contamination in estuaries using biological indicators. Mar Biol Assoc UK Occas Publ No. 4.

Bryan GW, Langston WJ (1992) Bioavailability, accumulation and effects of heavy metals in sediments with special reference to United Kingdom estuaries: a review. Environ Pollut 76:89–131.

Calabrese A, Collier RS, Nelson DA, MacInnes JR (1973) The toxicity of heavy metals to embryos of the American oyster *Crassostrea virginica*. Mar Biol 18: 162–166.

Calabrese A, Thurburg FP, Gould E (1977) Effects of cadmium, mercury and silver on marine animals. Mar Fish Rev 39:5–11.

Calabrese A, MacInnes JR, Nelson DA, Greig RA, Yevich PP (1984) Effects of long-term exposure to silver or copper on growth, bioaccumulation and histopathology in the blue mussel *Mytilus edulis*. Mar Environ Res 11:253–274.

Chao TT, Anderson BJ (1974) The scavenging of silver by manganese and iron oxides in stream sediments collected from two drainage areas of Colorado. Chem Geol 14:159–166.

Cherry RD, Heyraud M, Higgo JJW (1983) Polonium-210: its relative enrichment in the hepatopancreas of marine invertebrates. Mar Ecol Prog Ser 13:229–236.

Cowan CE, Jenne EA, Crecelius EA (1985) Silver speciation in seawater: the importance of sulfide and organic complexation. In: Sigleo AC, Hattori A (eds) Marine and Estuarine Geochemistry. Lewis Publishers, Chelsea, MI, pp 285–303.

Davis JA (1977) Adsorption of trace metals and complexing ligands at the oxide/water interface. Ph.D. dissertation, Stanford University, Stanford, CA.

Davis JA, Gunther AJ, O'Connor JM (1992) Priority pollutant loads from effluent discharges to the San Francisco Bay Estuary. Water Environ Res 64:134–140.

Dinnel PA, Stober QJ, Crumley SC, Nakatani RE (1982) Development of a sperm toxicity test for marine waters. In: Pearson JG, Foster RB, Bishop WE (eds) Aquatic Toxicology and Hazard Assessment: Fifth Conference ASTM STP766. American Society for Testing and Materials, Philadelphia, PA, pp 82–98.

Engel DW, Sunda WG, Fowler BA (1981) Factors affecting trace metal uptake and toxicity to estuarine organisms. I. Environmental parameters. In: Vernberg JF, Calabrese A, Thurberg FP, Vernberg WB (eds) Biological Monitoring of Marine Pollutants. Academic Press, New York, pp 127–144.

Erel Y, Patterson CC (1994) Leakage of industrial lead into the hydrocycle. Geochim Cosmochim Acta 58:3289–3296.

Eyster LS, Morse MP (1984) Development of the surf clam (*Spisula solidissima*) following exposure of gametes, embryos and larvae to silver. Arch Environ Contam Toxicol 13:641–646.

Flegal AR (1980) The geographic variation of silver in the black turban snail, *Tegula funebralis*. Environ Int 3:303–305.

Flegal AR, Coale KH (1989) Discussion: Trends in lead concentrations in major U.S. rivers and their relation to historical changes in gasoline-lead consumption. Water Res Bull 25: 1275–1277.

Flegal AR, Nriagu JO, Niemeyer S, Coale KH (1989) Isotopic tracers of lead contamination in the Great Lakes. Nature 339:455–458.

Flegal AR, Sañudo-Wilhelmy SA (1993) Comparable levels of trace metal contamination in two semi-enclosed embayments: San Diego Bay and South San Francisco Bay. Environ Sci Technol 27:1934–1936.

Flegal AR, Sañudo-Wilhelmy SA, Scelfo GM (1995) Silver in the eastern Atlantic Ocean. Mar Chem 49:315–320.

Goldberg ED, Koide M, Hodge V, Flegal AR, Martin JH (1983) U.S. Mussel Watch: 1977–1978 results on trace metals and radionuclides. Estuarine Coastal Mar Sci 16:69–93.

Hornberger MI, Luoma SN, van Geen A, Fuller C, Anima R (1996) Historical trends of trace metals in sediments of San Francisco Bay, California (unpublished manuscript).

Jenne EA, Luoma S (1977) The forms of trace elements in soils, sediments, and associated waters: an overview of their determination and bioavailability. In: Wildung RE, Drucker H (eds) Biological implications of metals in the environment. CONF-750929, National Technical Information Service, Springfield, VA, pp 110–143.

Jenne EA, Girvin DC, Ball JW, Blanchard JM (1978) Inorganic speciation of silver in natural waters – Fresh to marine. In: Klein DP (ed) Environmental Impacts of Artificial Ice Nucleating Agents. Dowden, Hutchinson, and Ross, Stroudsberg, PA, pp 41–61.

Johansson C, Cain DJ, Luoma SN (1986) Variability in the fractionation of Cu, Ag, and Zn among cytosolic proteins in the bivalve *Macoma balthica*. Mar Ecol Prog Ser 28:87–97.

Kuwabara JS, Luther GW III (1993) Dissolved sulfides in the oxic water column of San Francisco Bay, California. Estuaries 16:567–573.

Luoma SN, Bryan GW (1978) Factors controlling the availability of sediment-bound lead to the estuarine bivalve *Scorbicularia plana*. J Mar Biol Assoc UK 58:793–802.

Luoma SN, Bryan GW (1981) A statistical assessment of the form of trace metals in oxidized surface sediments employing chemical extractants. Sci Total Environ 17:165–196.

Luoma SN, Bryan GW (1982) A statistical study of environmental factors controlling concentrations of heavy metals in the burrowing bivalve *Scorbicula plana* and the polychaete *Nereis diversicolor*. Estuarine Coastal Shelf Sci 15:95–108.

Luoma SN, Cain DJ, Johansson CE (1985) Temporal fluctuations of silver, copper and zinc in the bivalve *Macoma balthica* at five stations in South San Francisco Bay. Hydrobiologia 129:413–425.

Luoma SN, Phillips DJH (1988) Distribution, variability, and impacts of trace elements in San Francisco Bay. Mar Pollut Bull 19:413–425.

Luoma SN, Dagovitz R, Axtmann E (1990) Temporally intensive study of trace metals in sediments and bivalves from a large river-estuarine system: Suisun Bay/ Delta in San Francisco Bay. Sci Total Environ 97/98:685–712.

Luoma SN, Ho YB, Bryan GW (1995) Fate, bioavailability and toxicity of silver in estuarine environments. Mar Pollut Bull 31:44–54.

Lussier SM, Cardin JA (1985) Results of acute toxicity tests conducted with silver at ERL, Narragansett. In: Ambient aquatic life water quality criteria for silver. U.S. Environmental Protection Agency, RI (unpublished manuscript).

Lyons WB, Fitzgerald WF (1983) Trace metals speciation in nearshore anoxic and suboxic pore waters. In: Wong CS, Bruland KW, Burton JD, Goldberg ED (eds) Trace Elements in Sea Water. Plenum Press, New York, pp 621–641.

Martin JH, Knauer GA, Gordon RM (1983) Silver distributions and fluxes in northeast Pacific waters. Nature 305:306–309.

Martin M, Ichikawa G, Goetzl J, de los Reyes M, Stephenson M (1984) Relationship between stress and toxic substances in the bay mussel, *Mytilus edulis*, from San Francisco Bay. Mar Environ Res 11:91–110.

Martin M, Stephenson MD, Smith DR, Guttierrez-Galindo EA, Florez-Muñoz G (1988) The use of silver in mussels as a tracer of domestic wastewater discharge. Mar Pollut Bull 19:512–520.

Measures CI (1995) The distribution of Al in the IOC stations of the eastern Atlantic between 30° S and 34° N. Mar Chem 49:267–281.

Miller LA, Bruland KW (1995) Organic speciation of silver in marine waters. Environ Sci Technol 29:2616–2621.

Murozumi M (1981) Isotope dilution surface ionization mass spectrometry of trace constituents in the Pacific. Bunseki Kagaku 30:19–36.

Nelson DA, Calabrese A, Greig R, Yevich PP, Chang S (1983) Long-term silver effects on the marine gastropod *Crepidula fornicata*. Mar Ecol Prog Ser 12:155–165.

O'Connor TP (1992) Mussel watch: recent trends in coastal environmental quality. National Oceanic and Atmospheric Administration (NOAA), Rockville, MD.

Patterson CC (1971) Native copper, silver, and gold accessible to early metallurgists. Am Antiquity 36:286–321.

Patterson CC, Settle DM (1976) The reduction of orders of magnitude errors in lead analyses of biological materials and natural waters by eliminating and controlling the extent and sources of industrial lead contamination introduced during sample collecting, handling, and analysis. In: LaFleur PD (ed) Accuracy in Trace Analysis: Sampling, Sample Handling, and Analysis. Spec Publ 422, National Bureau of Standards, U.S. Government Printing Office, Washington, DC, pp 321–351.

Patterson CC (1978) Silver stocks and losses in ancient and medieval times. Econ Hist Rev 25:205–235.

Rivera-Duarte I, Flegal AR (1996) Porewater silver concentration gradients and benthic fluxes from contaminated sediments of San Francisco Bay. Mar Chem (in press).

Rutherford F, Church TM (1975) The use of the metals silver and zinc to trace sewage sludge dispersal in coastal waters. In: Church TM (ed) Marine Chemistry in the Coastal Environment. American Chemical Society, Washington, DC, pp 440–452.

Sanders JG, Abbe GR (1987) The role of suspended sediments and phytoplankton in the partitioning and transport of silver in estuaries. Cont Shelf Res 7:1357–1361.

Sanders JG, Abbe GR, Riedel GF (1990) Silver uptake and subsequent effects on growth and species composition in an estuarine community. Sci Total Environ 97/98:762–769.

Sañudo-Wilhelmy SA, Flegal AR (1992) Anthropogenic silver in the Southern California Bight: a new tracer of sewage in coastal waters. Environ Sci Technol 26: 2147–2151.

Sañudo-Wilhelmy SA, Rivera-Duarte I, Flegal AR (1997) Distribution of colloidal trace metals in the San Francisco Bay estuary. Geochim Cosmochim Acta (in press).

Silver Institute (1994) World Silver Survey. The Silver Institute, 1112 Sixteenth Street, NW, Washington, DC.

Smith GJ, Flegal AR (1993) Silver in San Francisco Bay estuarine waters. Estuaries 16:547–558.

Smith IC, Carson BL (1977) Trace Metals in the Environment, Vol. 2. Ann Arbor Science, Ann Arbor, MI.

Soyer J (1963) Contribution to the study of the effects of mercury and silver on marine organisms. Vie Milieu 14:1–36.

Tanizaki Y, Shimokawa T, Nakamura M (1992) Physicochemical speciation of trace elements in river waters by size fractionation. Environ Sci Technol 26: 1433–1444.

Thomson EA, Luoma SN, Johansson CE, Cain DJ (1984) Comparison of sediments and organisms in identifying sources of biologically available trace metal contamination. Water Res 18:755–765.

Veron A, Church TM, Patterson CC, Flegal AR (1994) Distribution and transport of dissolved lead in North Atlantic surface waters. Geochim Cosmochim Acta 58:3199–3206.

Whitlow SI, Rice DL (1985) Silver complexation in river waters of Central New York. Water Res 19:619–626.

Wilson WB, Freeberg LR (1980) Toxicity of metals to marine phytoplankton cultures. EPA-600/3-870-025. National Technical Information Service, Springfield, VA.

Windom HL, Byrd JT, Smith RG Jr., Huan F (1991) Inadequacy of NASQAN data for assessing metal trends in the nation's rivers. Environ Sci Technol 25:1937–1941.

Zoto GA, Robinson WE (1985) The toxicity of one-hour silver exposures to early life stages of surf clam *Crassostrea virginica* and *Mytilus edulis*. Mar Ecol Prog Ser 12:167–173.

Manuscript received May 24, 1996; accepted June 1, 1996.

Index

Reviews of Environmental Contamination and Toxicology

Edited by

George W. Ware

Published by

Springer-Verlag New York • Berlin • Heidelberg • Barcelona • Budapest
Hong Kong • London • Milan • Paris • Santa Clara • Singapore • Tokyo

---

The original copy and one good photocopy of the manuscript, and a diskette with the electronic files for the manuscript, complete with figures and tables, are required. Manuscripts will be published in the order in which they are received, reviewed, and accepted. They should be sent to the editor:

Dr. George W. Ware
Department of Entomology
University of Arizona
Tucson, AZ 85721
Telephone and FAX: (520)299-3735
email: gware@ag.arizona.edu

1. Manuscript

The manuscript, in English, should be typewritten, double-spaced throughout (including reference section), on one side of 8½ × 11-inch blank white paper, with at least one-inch margins. The first page of the manuscript should start with the title of the manuscript, name(s) of author(s), with author affiliation(s) as first-page starred footnotes, and "Contents" section. Pages should be numbered consecutively in arabic numerals, including those bearing figures and tables only. In titles, in-text outline headings and subheadings, figure legends, and table headings only the initial word, proper names, and universally capitalized words should be capitalized.

Footnotes should be inserted in text and numbered consecutively in the text using arabic numerals.

Tables should be typed on separate sheets and numbered consecutively within the text in *arabic numerals*; they should bear a descriptive heading, in lower case, which is underscored with one line and starts after the word "Table" and the appropriate arabic numeral; *footnotes in tables* should be designated consecutively within a table by the lower-case alphabet. *Figures* (including photos, graphs, and line drawings) should be numbered consecutively within the text in arabic numerals; each figure should be affixed to a separate page bearing a legend (below the figure) in lower case starting with the term "Fig." and a number.

To facilitate production, authors are strongly encouraged to submit their manuscripts (including figures and tables) in electronic form on diskette. Manuscripts may be submitted in DOS, Windows, or Macintosh format (but not UNIX) using any popular word processing software (e.g., Word-Perfect, Microsoft Word) or they can be saved as an ASCII file. Tables can be prepared likewise or can be submitted as spreadsheets (e.g., Lotus 1-2-3, Microsoft Excel). Figures may also be submitted electronically using such programs as Adobe Illustrator, CorelDraw, MacDraw, and Aldus Superpaint. Figures should be saved both in their original application and as PostScript files. Authors with questions regarding electronic preparation of their manuscripts are encouraged to contact Hal Henglein at Springer-Verlag via phone (212-460-1546), FAX (212-533-5977), or Internet (HALH@SPRINGER-NY.COM).

2. Summary

A concise but informative summary (double-spaced) must conclude the text of each manuscript; it should summarize the significant content and major conclusions presented. It must not be longer than two $8\frac{1}{2} \times 11$-inch pages of double-spaced typing. As a summary, it should be more informative than the usual abstract.

3. References

All papers, books, and other works cited in the text must be included in a "References" section (*also double spaced*) at the end of the manuscript. If comprehensive papers on the same subject have been published, they should be cited when the bibliographic citations extend farther back than to these papers.

All papers cited in the text should be given in parentheses and alphabetically when more than one reference is cited at a time, e.g. (Coats and Smith 1979; Holcombe et al. 1982; Stratton 1986), except when the author is mentioned, as for example, "and the study of Roberts and Stoydin (1985)." References to unpublished works should be kept to a minimum and mentioned only in the text itself in parentheses. References to published works are given at the end of the text in alphabetical order under the first author's name and chronologically, citing all authors (surnames followed by initials throughout; do not use "and") according to the following examples:

Periodicals: Name(s), initials, year of publication in parentheses, full article title, journal title as abbreviated in "The ACS Style Guide: A Manual for Authors and Editors" of the American Chemical Society, volume number, colon, first and last page numbers. Example:

Leistra MT (1970) Distribution of 1,3-dichloropropene over the phases in soil. J Agric Food Chem 18:1124–1126.

Books: Name(s), initials, year of publication in parentheses, full title, edition, volume number, name of publisher, place of publication, first and last page numbers. Example:

Gosselin R, Hodge H, Smith R, Gleason M (1976) Clinical Toxicology of Commercial Products, 4th ed. Wilkins-Williams, Baltimore, MD, pp 119–121.

Work in an edited collection: Name(s), initials, year of publication in parentheses, full title. In: name(s) and initial(s) of editor(s), the abbreviation ed(s) in parentheses, name of publisher, place of publication, first and last page numbers. Example:

Metcalf RL (1978) Fumigants. In: White-Stevens J (ed) Pesticides in the environment. Marcel Dekker, New York, pp 120–130.

Abbreviations

A	acre	min	minute(s)
bp	boiling point	$\underline{M}$	molar
cal	calorie	mon	month(s)
cm	centimeter(s)	ng	nanogram(s)
d	day	nm	nanometer(s) (millimicron)
ft	foot (feet)	$\underline{N}$	normal
gal	gallon(s)	no.	number(s)
g	gram(s)	od	outside diameter
ha	hectare	oz	ounce(s)
hr	hour(s)	ppb	parts per billion (μg/kg)
in.	inch(es)	ppm	parts per million (mg/kg)
id	inside diameter	ppt	parts per trillion (ng/kg)
kg	kilogram(s)	pg	picogram
L	liter(s)	lb	pound(s)
mp	melting point	psi	pounds per square inch
m	meter(s)	rpm	revolutions per minute
m^3	cubic meter	sec	second(s)
μg	microgram(s)	sp gr	specific gravity
μL	microliter(s)	sq	square (as in "sq m")
μm	micrometer(s)	vs	versus
mg	milligram(s)	wk	week(s)
mL	milliliter(s)	wt	weight
mm	millimeter(s)	yr	year(s)
m$\underline{M}$	millimolar		

Numbers: All numbers used with abbreviations and fractions or decimals are arabic numerals. Otherwise, numbers below ten are to be written out. Numerals should be used for a series (e.g., "0.5, 1, 5, 10, and 20 days"), for pH values, and for temperatures. When a sentence begins with a number, write it out.

Symbols: Special symbols (e.g., Greek letters) must be identified in the margin, e.g.

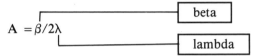

Percent should be % in text, figures, and tables.

Style and format: The following examples illustrate the style and format to be followed (except for abandonment of periods with abbreviation):

Sklarew DS, Girvin DC (1986) Attenuation of polychlorinated biphenyls in soils. Reviews Environ Contam Toxicol 98:1–41.
Yang RHS (1986) The toxicology of methyl ethyl ketone. Residue Reviews 97:19–35.

References by the same author(s) are arranged chronologically. If more than one reference by the same author(s) published in the same year is cited, use a, b, c after year of publication in both text and reference list.

4. Illustrations

Illustrations may be included only when indispensable for the comprehension of text. They should not be used in place of concise explanations in text. Schematic line drawings must be drawn carefully. For other illustrations, clearly defined black-and-white glossy photos are required. Should darts (arrows) or letters be required on a photo or other type of illustration, they should be marked neatly with a soft pencil on a duplicate copy or on an overlay, with the end of each dart indicated by a fine pinprick; darts and lettering will be transferred to the illustrations by the publisher.

Photos should not be less than 5 × 7 inches in size. Alterations of photos in page proof stage are not permitted. *Each photograph or other illustration should be marked on the back, distinctly but lightly, with soft pencil, with first author's name, figure number, manuscript page number, and the side that is the top.*

If illustrations from published books or periodicals are used, the exact source of each should be included in the figure legend: if these "borrowed" illustrations are copyrighted by others, permission of the copyright holder to reproduce the illustrations must be secured by the author. Permissions forms are available from the Editor and upon completion by the original publisher should be returned to Janet Slobodien, Life Sciences Editorial, Springer-Verlag, 175 Fifth Avenue, New York, NY 10010.

5. Chemical Nomenclature

All pesticides and other subject-matter chemicals should be identified according to *Chemical Abstracts*, with the full chemical name in text in parentheses or brackets the first time a common or trade name is used. *If many such names are used, a table of the names, their precise chemical designations, and their* Chemical Abstract Numbers (CAS) *should be included as the last table in the manuscript, with a numbered footnote reference to this fact on the first text page of the manuscript.*

6. Miscellaneous

Abbreviations: Common units of measurement and other commonly abbreviated terms and designations should be abbreviated as listed below; if any others are used often in a manuscript, they should be written out the first time used, followed by the normal and acceptable abbreviation in parentheses [e.g., Acceptable Daily Intake (ADI), Angstrom (Å), picogram (pg)]. Except for inch (in.) and number (no., when followed by a numeral), abbreviations are used without periods. Temperatures should be reported as "°C" or "°F" (e.g., mp 41° to 43°C). Because the metric system is the international standard, when pounds (lb) and gallons (gal) are used the metric equivalent should follow in parentheses.

7. Proofreading scheme

The senior author must return the Master set of page proofs to Springer-Verlag within one week of receipt. Author corrections should be clearly indicated on the proofs with ink, and in conformity with the standard "Proofreader's Marks" accompanying each set of proofs. In correcting proofs, new or changed words or phrases should be carefully and legibly handprinted (not handwritten) in the margins.

8. Offprints

Senior authors receive 30 complimentary offprints of a published paper. Additional offprints may be ordered from the publisher at the time the principal author receives the proofs. Order forms for additional offprints will be sent to the senior author along with the page proofs.

9. Page charges

There are no page charges, regardless of length of manuscript. However, the cost of alteration (other than corrections of typesetting errors) attributable to authors' changes in the page proof, in excess of 10% of the original composition cost, will be charged to the authors.

If there are further questions, see any volume of *Reviews of Environmental Contamination and Toxicology* (formerly *Residue Reviews*) or telephone the Editor (see first page for telephone number). Volume 143 is especially helpful for style and format.